AF364435

Objective
Soil Science

Objective
Soil Science

M. Raghavendra Reddy

P.N. Siva Prasad

A. Sathish

V.R. Ramakrishna Parama

BS Publications
A unit of **BSP Books Pvt. Ltd.**
4-4-309/316, Giriraj Lane, Sultan Bazar,
Hyderabad - 500 095.

Objective Soil Science
by M. Raghavendra Reddy, P.N. Siva Prasad, A. Sathish and V.R. Ramakrishna Parama

© 2019, *by Publisher*

Published by:

BS Publications
A unit of **BSP Books Pvt., Ltd.**

4-4-309/316, Giriraj Lane, Sultan Bazar,
Hyderabad - 500 095
Phone : 040 - 23445600, 23445688
e-mail : info@bspbooks.net
www.bspbooks.net

ISBN: 978-93-88305-94-5 (Hardback)

Preface

Soil science and Agricultural chemistry is one of the major subjects in Agricultural sciences. It became a deciding subject to become candidature for most of the aspirants for most competitive exams in Agriculture.

This book contains all major divisions of the subject, comprising the various formats of objective questions which are frequently found in different examinations. In addition, very important corners of subject to be recalled for various examinations are included in this book.

Applied and supporting subjects like Remote sensing and Nanotechnology are also included in this book to update and prepared for future exams.

This book will be a great asset to aspirants for Junior Research Fellow (JRF) and Senior Research Fellow (SRF) conducting by Indian Council of Agricultural Research (ICAR) and other competitive exams.

We hope that this book is a valuable resource for all aspirants to achieve a great career. We will be happy with your suggestions for enhancing the effectiveness of the text and other suggestions to improve the quality of book effectively.

M. Raghavendra Reddy
P.N. Siva Prasad
A. Sathish
V.R. Ramakrishna Parama

Acknowledgements

I am indebted to all authors contributed various chapters in this book, professors, scholars and student friends of the Department of Soil Science and Agricultural Chemistry, GKVK, Bengaluru who fulfilled my wish.

I express my deep sense of gratitude and sincere thanks to Dr. V. R. Ramakrishna Parama, Professor (Rtd.), Department of Soil Science and Agricultural Chemistry for his encouragement, constructive criticism and useful comments without which this book would not have reached this stage.

With heartfelt gratefulness, I express my profound reverence to my beloved parents for their love, affection and support.

M. Raghavendra Reddy
P.N. Siva Prasad
A. Sathish
V.R. Ramakrishna Parama

Contents

<table>
<tr><td>**1**</td><td># SOIL PEDOLOGY</td></tr>
<tr><td></td><td>**M. Raghavendra Reddy and V.R.R. Parama**</td></tr>
</table>

CHOOSE THE CORRECT ANSWER

1. Smallest volume of that can be called a soil is _________________

   (a) Pedon   (b) Polypedon   (c) Horizon   (d) Series

2. Who enunciated for the first time the word "Fundamental principles of pedology"

   (a) Jenny   (b) Dockuchaiev   (c) Joffee   (d) Marbut

3. Who written the book "The great soil groups of the world"

   (a) Smith   (b) Vant Hoff

   (c) Schokalskaya   (d) Glinka

4. Thickness of earth crust is _____________

   (a) 5 to 11 km   (b) 5 to 11m

   (c) 60 to 600 km   (d) 60 to 600

5. Innermost position of earth is _________________

   (a) Crust   (b) Mantle

   (c) Core   (d) None of the above

6. Average composition of earths crust (% by weight) in order of $O^{2-} > Si^{4+} > Al^{3+} > \underline{\hspace{2cm}} > Ca^{2+}$

   (a) K^+ (b) Na^+ (c) Mg^{2+} (d) Fe^{2+}

7. Composition of the upper 5 km of earths crust consists a majority of _____________________ rocks (74%).

   (a) Igneous rocks   (b) Sedimentary rocks

   (c) Metamorphic rocks   (d) (a+c)

8. As a whole, earth crust consists majority of _____________ rocks and _____________________ percentage

   (a) Igneous, 95 %   (b) Sedimentary, 95 %

   (c) Igneous, 74%   (d) Sedimentary, 74 %

9. Cleavage is absent in _____________ among the below minerals

   (a) Pyroxene   (b) Amphibole   (c) Both   (d) Olivine

10. Find the Ferromagnesian mineral
    (a) Biotite   (b) Muscovite   (c) Orthoclase (d) a & b

11. Gibbsite formula
    (a) $[Al_3(OH)_4]_n$ (b) $[Mg_3(OH)_4]_n$
    (c) $[Al_2(OH_6)]_n$ (d) $[Mg_3(OH)_6]_n$

12. Attapulgite & polygorskite is mostly found in region of ______________
    (a) Humid   (b) Temperate   (c) Arid   (d) All region

13. Among the following, which is not primary mineral
    (a) Quartz   (b) Serpentine   (c) Mica   (d) Hornblende

14. Cooling and crystallization of magma forms ______________ rocks
    (a) Igneous   (b) Sedimentary
    (c) Metamorphic   (d) All of the above

15. Transformation of unconsolidated sediments to hard rocks is ______________
    (a) Weathering   (b) Transportation
    (c) Deposition   (d) Diagenesis

16. Banded or laminated character is the most peculiar feature of ______________ rocks
    (a) Igneous   (b) Metamorphic
    (c) Sedimentary   (d) All of these

17. Weathering is ______________
    (a) Disintegration   (b) Decomposition
    (c) Both a & b   (d) None of the above

18. Exfoliation (or) onion type of weathering is due to changes in ______________
    (a) Temperature   (b) Pressure
    (c) Oxygen   (d) Wind blast

19. Rill formation in limestone is due to ______________
    (a) Reduction   (b) Carbonation
    (c) Oxidation   (d) Hydration

20. Weathering of quartz (desilication) is favored by ______________
    (a) High temperature   (b) Intense leaching
    (c) Parent material   (d) All of these

21. Among the following, which weathers fastly _____________

    (a) Feldspars   (b) Quartz   (c) Gibbsite   (d) Haematite

22. B-Horizon formation may absent in _____________ condition

    (a) Cool & humid   (b) Warm & Humid

    (c) Cool & arid   (d) Warm & arid

23. Soil forming factors are divided into active and passive factors by

    ____

    (a) Joffee   (b) Jenney

    (c) Dockuchaiev   (d) Marbut

24. Red colour of soils are due to _____________

    (a) Si-Al   (b) Ca & Mg   (c) Na & K   (d) Fe & Mn

25. The term _____________ is used for a sequence of topographically related soils which have comparable parent material, climate and age but show different characteristics owing to variation in relief and drainage

    (a) Master horizon (b) Catena   (c) Profile   (d) Series

26. Five stages of weathering are given by _____________

    (a) Milne   (b) Tripathi

    (c) Mohr & Van Baren   (d) Klingebiel

27. Young alluvial soils are also called _____________

    (a) Bhangar   (b) Khadar

    (c) Both   (d) None of the above

28. Among the all soil forming factors, the most predominant is _____________

    (a) Parent material   (b) Relief

    (c) Climate   (d) All of the above

29. Production and addition of organic matter is more in _____________

    (a) Grass soils   (b) Forest soils

    (c) Laterite soils   (d) Dry land soils

30. Forest soil horizon (Ah), is of intimately - mixed mineral matter and amorphous humus is _____________

    (a) Mor   (b) Mull   (c) Sward   (d) Orterde

31. Termites work as soil forming agents at any time in _____________ climate

    (a) Arid   (b) Humid   (c) Both   (d) It wont work

32. Highly - decomposed organic matter in ____________ horizon

    (a) Ah   (b) Ap   (c) Oa   (d) Oe

33. Among the following, which is not fundamental pedogenic

    (a) Humification   (b) Illuviation

    (c) Alkalization   (d) Eluviation

34. Eluviation is most pronounced in ________________ soils

    (a) Podsols   (b) Laterite soils

    (c) Lateritic soils   (d) Mollisols

35. Zone of illuviation is ________________

    (a) O horizon   (b) A horizon

    (c) B horizon   (d) C horizon

36. Most important process of chemical weathering is ____________

    (a) Hydrolysis   (b) Hydration

    (c) Solution   (d) All the above

37. Podzolization is negative process of ________________

    (a) Pedoturbation   (b) Calcification

    (c) Gleization   (d) Salinization

38. Find the odd, unrelated soils among following ________________

    (a) Lato soils   (b) Laterite soils

    (c) Oxisols   (d) Andisols

39. Very deeply weathered soils are ________________

    (a) Red soils   (b) Black soils

    (c) Laterite soils   (d) Alluvial soils

40. Soil forming process under water logged soils is ________________

    (a) Gleization   (b) Podsolization

    (c) Calcifiation   (d) Decalification

41. Which of the following soil forming process is not zonal soil forming process ________________

    (a) Calcification   (b) Podzolization

    (c) Gleization   (d) Laterization

42. Salinization and solonization brings ____________ & ________________ soil structure

    (a) Columnar & granular   (b) Granular & Columnar

    (c) Honey comb & Granular   (d) Granular & Honey comb

43. _______________ is a group of soils, similar in their differentiating characteristics and arrangement of horizons

(a) Soil series
(b) Soil phase
(c) Soil family
(d) Master horizon

44. Soil survey for agricultural research stations is _______________

(a) Detailed
(b) Semi-detailed
(c) Reconnassonce
(d) Detailed- reconnassonce

45. Which of the following scale is comes under cadestral map _______________

(a) 1: 250,000
(b) 1:25,000
(c) 1:8000
(d) All of these

46. Area represented by 1 cm^2 on map in reconnaissance survey with scale of 1:100,000 is __________ha

(a) 10
(b) 100
(c) 1000
(d) 10000

47. Frequency of observations needed for high intensity detailed survey for a village is _______________ per hectare

(a) 2
(b) 3
(c) 4
(d) 1

48. In the 3 tier approach of soil resource mapping the 1st tier is _______________

(a) Image interpretation
(b) Field survey
(c) Laboratory analysis
(d) Cartography

49. Grouping of soils based on their productivity is _______________ classification

(a) Physical
(b) Chemical
(c) Geological
(d) Economic

50. Alluvial soils are _______________

(a) Zonal
(b) Intrazonal
(c) Azonal
(d) None of these

51. Accumulation of iron and Aluminium show characters of _______________

(a) Pedocals
(b) Pedalfers
(c) Both
(d) None of these

52. Developed profiles can see in _______________ soils

(a) Zonal soils
(b) Intra zonal soils
(c) Azonal soils
(d) All of these

53. Number of orders in morpho-genetic system of soil classification

(a) Zero
(b) One
(c) Two
(d) Three

54. Number of epipedons are ______________
    (a) 9   (b) 10   (c) 11   (d) 8

55. Surface horizon, having mollic properties but BS is less than 50% is
    ____

    (a) Melanic   (b) Umbric   (c) Anthropic (d) Ochric

56. Grossarenic epipedon is a horizon with ______________ cm or more thick
    over an argillic horizon
    (a) 120   (b) 80   (c) 100   (d) 150

57. Organic horizon with very high organic matter, remains saturated with
    water for 30 or more days (cumulative) is ________________
    (a) Histic   (b) Mollic   (c) Melanic   (d) Umbric

58. Columnar or prismatic structure can seen in ______________ horizon
    (a) Argillic   (b) Natric   (c) Spodic   (d) Kandic

59. Extremely variable horizon in minerology because of its pedogenic youth-
    fulness is ________________
    (a) Sombic   (b) Oxic   (c) Cambic   (d) Albic

60. A sub surface horizon, at least half cemented by SiO_2 is
    (a) Durinode   (b) Duripan   (c) Fragipan   (d) Plinthite

61. A humus poor, sesquioxide - rich horizon is ________________
    (a) Durinode   (b) Duripan   (c) Fragipan   (d) Plinthite

62. Tonguing is common in ______________ horizon
    (a) Glossic   (b) Grossarenic   (c) Gypsic   (d) Petrogypsic

63. Peat soils are having ________________ soil material
    (a) Hemic   (b) Fibric   (c) Sapric   (d) Humilluvic

64. Depth of soil moisture control section is more for
    ________________ soils
    (a) Fine   (b) Medium   (c) Coarse   (d) Same for all

65. Which soil moisture regime is covered most of Indian soils
    (a) Aridic   (b) Udic   (c) Xeric   (d) Ustic

66. Poorly drained soils are having ______________ region
    (a) Aridic   (b) Ustic   (c) Udic   (d) Aquic

67. Morphogenetic approach of soil classification was given by
    ____
    (a) Marbut   (b) Baldwin   (c) Kellogg   (d) Throp

68. Every 10 cm increase in depth diurnal fluctuations changes in soil temperature by _______ ^{0} c

(a) 0.5   (b) 0.6   (c) 0.8   (d) 1

69. Mean annual soil temperature of thermic region is ___________ ^{0}C

(a) 28 to 35   (b) 22 to 28   (c) 8-15   (d) 15 to 22

70. Which of the following is not tropical soil temperature regime _____________

(a) Cryic   (b) Mesic   (c) Thermic   (d) Mega thermic

71. C_3 pulse crops are needs _______________ soil temperature regime

(a) Hyperthermic   (b) Mega thermic   (c) Cryic   (d) Frigid

72. Maximum area of Indian soils are falls under _________________ soil temperature regime

(a) Hyperthermic   (b) Thermic

(c) Megathermic   (d) Cryic

73. The highest category in soil classification is ____________

(a) Order   (b) Suborder

(c) Great group   (d) Phase

74. On 7th approximation soil classification, category which is based on diagnostic subsurface horizons

(a) Suborder   (b) Sub group

(c) Great group   (d) Family

75. Number of soil orders according to 7th approximation _______________

(a) 12   (b) 10   (c) 14   (d) 16

76. Argillic horizon with high base saturation can found in

(a) Ultisols   (b) Aridsoils   (c) Alfisols   (d) Oxisols

77. Distinct horizons are absent in _____________ order

(a) Aridisols   (b) Inceptisol   (c) Entisols   (d) Vertisols

78. Laterite soils are categorised in __________________ order

(a) Alfisols   (b) Aridisols   (c) Vertisols   (d) Oxisols

79. Azonal soils are categorized in __________________ order

(a) Entisols   (b) Inceptisols   (c) Alfisols   (d) Oxisols

80. Soils of grassland vegetation classified in __________________ order

(a) Histosols   (b) Mollisols   (c) Spodosols   (d) Entisols

81. Recently formed soils may have ________________ epipedon
    (a) Ochric   (b) Anthropic   (c) Both   (d) Umbric

82. Low activity clays can seen in ____________ endopedon
    (a) Kandic   (b) Octric   (c) Anthropic   (d) Umbric

83. Fluvents, a suborder of ______________
    (a) Oxisol   (b) Alfisol   (c) Vertisol   (d) Entisol

84. At early stage of soil formation, the soils formed will be __________ soil order
    (a) Entisol   (b) Vertisol   (c) Oxisol   (d) All of these

85. Argill - pedoturbation (haplodization) can seen in ________________
    (a) Alluvial soil   (b) Laterite soil   (c) Black soil   (d) Red soil

86. Imperfectly to poor drained soils can seen in ________________ order
    (a) Alfisols   (b) Andisols   (c) Oxisols   (d) Vertisols

87. Mollic epipedon can seen in ________________
    (a) Mollisols   (b) Vertisols   (c) Inceptisols   (d) All of these

88. The suborder ustolls is falls in ________________ soil order
    (a) Vertisols   (b) Mollisols   (c) Inceptisols   (d) Ultisols

89. ________________ are mineral soils of dry places and of areas having high ground water table
    (a) Aridisols   (b) Ultisols   (c) Oxisol   (d) All of these

90. Bleached E-horizon is a character of ____________ soil order
    (a) Andisol   (b) Gelisol   (c) Spodosol   (d) Entisol

91. Aqods & Orthods are suborder of ________________ soil order
    (a) Spodosol   (b) Entisol
    (c) Histosols   (d) None of these

92. Oxisols are ______________ in soil fertility
    (a) Good   (b) Very good
    (c) Poor   (d) Very poor

93. Oxisols can found in ____________ state/ s of India
    (a) Kerala   (b) Karnataka
    (c) Tamilnadu   (d) All of these

94. ________________ soil order is being acidic, face the problem of phosphorus fixation by Fe & Al
    (a) Vertisols   (b) Entisols
    (c) Oxisols   (d) All of these

95. To classify a soil as Histosol, the OM (%) should be minimum of
____

   (a) 15   (b) 10   (c) 12   (d) 20

96. Soil order wheels is not found in India

   (a) Gelisol   (b) Andisol

   (c) Spodosols   (d) All of these

97. Peat, Muck & bog soils are classified to

   (a) Mollisols   (b) Histosols

   (c) Spodosols   (d) All of these

98. Most recently introduced order in 7th approximation is ___________

   (a) Gelisol   (b) Andisol

   (c) Histosol   (d) Inceptisol

99. Degree of weathering is maximum for ___________

   (a) Mollisol   (b) Spodosol

   (c) Oxisol   (d) Inceptisol

100. Accuracy and reliability of prediction of a soil is more at ___________ level of taxonomy

   (a) Order   (b) Great group

   (c) Family   (d) Series

101. ___________ group of soils having a natric horizon

   (a) Solonetz   (b) Solonchacks

   (c) Nitisols   (d) Regosols

102. Most of the Alfisols are ___________ in colour

   (a) Black   (b) Red   (c) Both   (d) White

103. Agroecological region are classified based on ___________

   (a) Length of growing period   (b) Soil

   (c) Bioclimatic conditions   (d) All of these

104. Number of Agroecological regions according to NBSS & LUP ___________

   (a) 20   (b) 24   (c) 25   (d) 15

105. Climatic zone having LGP more than 210 days is ___________

   (a) Sub Humid   (b) Per humid   (c) Semiarid   (d) Arid

106. Soil order predominant in India is ___________

   (a) Vertisol   (b) Alfisols

   (c) Entisols   (d) Inceptisols

107. Alluvial soils are deficient in ________________

    (a) Nitrogen   (b) Phosphorus

    (c) Humus   (d) All of these

108. Which of the following is not a character of black soil

    (a) Good aeration   (b) Vertical mulching

    (c) Churning   (d) Swell & Shrink

109. Growing period of desert soils is ___________ days

    (a) > 70   (b) > 90   (c) < 60   (d) < 100

110. Red soils of Inida, mostly develops on _____________ rocks

    (a) Acidic   (b) Basic   (c) Neutral   (d) Calcareous

111. Silica: Sesquioxide ratio of red soils is ___________________

    (a) < 1   (b) < 2.5   (c) 2.5 to 3.0   (d) > 3.0

112. Cation exchange capacity is less in _____________ soils

    (a) Red   (b) Black   (c) Alluvial   (d) Both a & b

113. Hue representing 7.5 year is characteristics of ___________ soils

    (a) Black   (b) Red

    (c) Alluvial   (d) All of these

114. Highly weathered material enriched in secondary forms of iron and aluminium in _____________ soils

    (a) Laterite   (b) Red   (c) Black   (d) Alluvial

115. Low activity clays are found in __________ soils

    (a) Red   (b) Laterite   (c) Black   (d) Alluvial

116. Brick like soil material is characteristic of ____________ soils

    (a) Red   (b) Black   (c) Desert   (d) Laterite

117. Majority of soil degradation is through _______________

    (a) Wind erosion   (b) Water erosion

    (c) Chemical deterioration   (d) Physical deterioration

118. ________________ is general appearance of a mineral in a reflected light

    (a) Colour   (b) Streak   (c) Lustre   (d) Tenacity

119. Hardness is more for ______________

    (a) Diamond   (b) Quartz   (c) Apatite   (d) Talc

120. Degree of penetration of tight through a mineral is its _______________

    (a) Lustre   (b) Tenacity

    (c) Streak   (d) Transparency

121. Quartz is _______________ in its transparency

    (a) Transparent   (b) Opaque

    (c) Translucent   (d) None of these

122. The ratio between the weight of a mineral or a substance to the weight of an equal volume of water

    (a) Cleavage   (b) Specific gravity

    (c) Striation   (d) Lustre

123. The property of a mineral to break along an irregular surface is _______________

    (a) Cleavage   (b) Breakage

    (c) Fracture   (d) All of these

124. Which of the following rock is a volcanic rock _______________

    (a) Gabbro   (b) Diorite   (c) Granite   (d) Basalt

125. CEC of vermiculite will be _______________ C mol (P^+) kg^{-1} clay

    (a) 120   (b) 30   (c) 10   (d) 200

126. Soil colour denoting by _______________

    (a) Hue   (b) Value

    (c) Chroma   (d) All of these

127. 8 YR 3/1 is the colour of soil, the number 3 denotes _______________

    (a) Hue   (b) Value

    (c) Chroma   (d) None of these

128. Chemical weathering is a decomposition process which takes place primarily at the junction of lithosphere and atmosphere called _______________

    (a) Weathering front   (b) Weathering back

    (c) Weathering down   (d) Weathering junction

129. _______________ soils develop a normal profile under the influence of active soil forming factors

    (a) Endodynamorphic soils   (b) Ectodynamorphic soils

    (c) Both   (d) None of these

130. Which of the epipedon is not present India
     (a) Mollic   (b) Ochric   (c) Umbric   (d) Folistic

131. Which level of the data give maximum information
     (a) 4th order   (b) 3rd order   (c) 2nd order   (d) 1st order

132. Which of the following is not a 2:1 type expanding mineral
     (a) Beidellite   (b) Nontronite
     (c) Montmorillonite   (d) Vermiculite

133. Shape of Kaolimite minerals is ___________________
     (a) Hexagonal plates   (b) Elongate
     (c) Irregular flakes   (d) Shapeless

134. Low using in X - ray diffraction method
     (a) Ficks law   (b) Bernoulies law
     (c) Miller law   (d) Braggs law

135. Inert material used in DTA (Differential Thermal Analysis) is ___________
     (a) Attapulgite   (b) Al_2O_3
     (c) Fe_2O_3 (d) $Al_2(OH)_6$

136. Y - Index of Illite type clays is in range of _______________
     (a) 5 to 25   (b) 0 to 5   (c) 25-35   (d) > 35

137. Physical downward movement of clay minerals is ________________
     (a) Melanization   (b) Mineralization
     (c) Lessivage   (d) Neutralization

138. Book namely "Benchmark soils of India" written by ___________________
     (a) R.S.Murthy   (b) K.L.Murthy
     (c) S.C.Yadv   (d) Ramamoorthy

139. Cryoturbation can seen in ___________________
     (a) Aridisols   (b) Inceptisols   (c) Mollisols   (d) Gelisols

140. The soil process associated with the formation of Histosols is ________________
     (a) Paludization   (b) Melanization
     (c) Leucinization   (d) Rubifaction

141. Category used for making prediction of land use plans
     (a) Sub group   (b) Family   (c) Series   (d) Order

142. Volcanic ash soils are categorised to __________________
     (a) Oxisols   (b) Histosols   (c) Andisols   (d) Gelisols

143. Minimum organic carbon of Histosols, when parent clay is zero in organic carbon percentage ___________

    (a) 6   (b) 8   (c) 10   (d) 12

144. Vertical section of soil is ________________

    (a) Horizon   (b) Phrase   (c) Series   (d) Profile

145. Removal of silica and accumulation of sesquioxide is ______________ process

    (a) Laterization   (b) Podzolization

    (c) Calcification   (d) None of these

146. ____________ process is dominant in black cotton soils

    (a) Laterization   (b) Regur formation

    (c) Gleization   (d) Humification

147. Weathering index of gypsum is ____________________

    (a) 1   (b) 5   (c) 6   (d) 10

148. The statement "Orthoclare (albite)" is more stable then anorthite

    (a) True   (b) False

    (c) Equally stable   (d) Cant be predictable

149. ______________ is index mineral for advanced stage of weathering

    (a) Hematite   (b) Gibbsite   (c) Kaolinite   (d) Quartz

150. The term soil catena is given by __________

    (a) Glinka   (b) Milne   (c) Marbut   (d) R.S.Murthy

FILL IN THE BLANKS

1. ____________________ is the father of soil science.

2. ______________ is defined as loose, unconsolidated inorganic material (debris) of weathered rocks on the earths surface.

3. Rock alters to regolith by ______________ process and further alters to true soil by ________________________.

4. Average composition of oxygen in earths crust ______________.

5. Formula of gypsum __________________.

6. The minerals formed due to weathering of pre-existing primary minerals are ________________________________.

7. Phyllosilicates consists _________ number of unshared and ______________ shared oxygen.

8.   Light minerals should have specific gravity less than _______________.

9.   Rocks formed due to transportation and cementation are _______________.

10. Rocks formed by solidification of the magma within the earths crust are called _______________ rocks.

11. Volcanic rocks are _______________ grained

12. Rocks containing large amount (> 60 %) of silica are termed _______________ rocks.

13. _______________ rocks is very light weighted rocks that has a lower specific gravity than water.

14. Stratification is the most common feature of _______________ rocks.

15. Four stages of formation of sedimentary rocks are _______________, _______________, _______________ and _______________.

16. Unconsolidated residues of weathering or weathered seeks on the earths surface is _______________.

17. Process of removal or loss of oxygen is _______________.

18. _______________ process is mechanical movement of day and iron oxides from A-horizon without undergoing chemical alteration.

19. Climate requirement to podzolization is _______________.

20. Bleached A-horizon is appears in _______________.

21. Climate requirement for formation of laterite soils _______________.

22. _______________ process is refers to removal of Na^+ from the exchange sites.

23. In a grid survey, gradually _______________ observations per hectares are recommended.

24. A soil _______________ consists of aggregate of all soils delineations which are identified by unique symbol, color or name on map.

25. Zonality concept was given by _______________.

26. _______________ defines central concept of great groups.

27. Diagnostic surface horizons are called _______________.

28. Man made surface horizon is _______________ epipedon.

29. A silicate clay enriched sub surface horizon formed by illuviation clay is _______________.

30. An eluvial or bleached E- horizon of podzols is ___________________.

31. ___________________ are weekly connected by SiO_2 to indurated nodules with a diameter of 1 cm or more.

32. Sub soil layer with high bulk density and brittle when moist, very hard when dry is _______________.

33. Soil moisture regime at wilting point (> 15 bar) is ___________________.

34. Prefix "iso" is used if the difference between the mean summer and mean winter temperatures is less than _______ ^{0} (C)

35. "Typic" is used to define ___________________ of a great group in category of sub group.

36. *Hapludoll* is ____________ soil order.

37. ____________ soil moisture regime is partly dry with moist summer.

38. Formative element in name of Histosols is _______________.

39. Tundra soils are comes under _______________ order.

40. Volcanic - ash soils are comes under _______________ order.

41. Black cotton soils are comes under _______________ order.

42. Cracking, followed by filling of cracks and swelling, result in development of ___________________________.

43. Shrinking & swelling of clays can seen in ___________________ order.

44. The most important soil forming processes operative in aridisols are ___________________ and ___________________.

45. The sub order *cryids*, *argids*, *calcids* are belongs to _______________ order.

46. Soils of ___________________ order are with accumulation of sesquioxides and humus in subsurface horizons.

47. Ultisols are comparable with ___________________, except for having low base Saturation.

48. Ultisols are develop in ___________________ climatic conditions.

49. *Xerult & Udults* are sub order of _______________.

50. ___________________ soil order is chemically degraded soils.

51. Shifting cultivation is common practice in ____________ soil order.

52. Churning is most common is ____________ soils

53. Silica: Sesquioxide ratio of black soils will be _____________

54. The expansion and contraction leads to formation of micro - topographic features, known as ____________

55. Attapulgite is a typical mineral of ____________ soils

56. ____________________ soils have hue of 5 YR with maximum intensity in B horizon

57. Silica : Sesquioxide ratio of laterite soils is ______________

58. Expand GLASOD ___________________

59. Iron minerals have ____________ lustre

60. The resistance of a mineral to scratching is known as its ______________

61. Tendency of a mineral to split in certain preferred directions along smooth plane surfaces is called ___________

62. The specific atomic arrangement of the minerals (as in a crystal) is called its ________________________

63. ______________ denotes dominant spectral color of the soil

64. ______________ denotes lightness or darkness of a color of soil

65. ______________ represents the strength of soil color

66. ______________ is the study of a soil from the stand point of higher plant

67. According to Braggs law [$n\lambda = 2d \sin è$] where, d is ____________________

68. Martin & Russell proposed ___________ for identification of minerals

69. Cadastral maps used in ______________ type of survey

70. Light - weight rock that has a lower specific gravity than water is _________

MATCH THE FOLLOWING

1.

A (Mineral)		B (Colour)	
1	Olivine	a	Pink (flesh)
2	Pyroxene	b	Drark - Green
3	Quartz	c	Olive - Green
4	Biotite	d	Colourless to grey
5	Orthoclare	e	Black

2.

A (Original rock)			B (Metamorphosed rock)	
1	Conglomerate		a	Marble
2	Sand stone		b	Mica- schist
3	Lime stone		c	Gneiss
4	Gabbro		d	Quartz - schist
5	Granite		e	Chlorite - schist

3.

A (Parent material)			B (Deposited by)	
1	Alluvium		a	Gravity
2	Colluvium		b	Lake
3	Lacustrine		c	Ice
4	Marine		d	River
5	Moraine		e	Ocean

4.

A (Soils)			B (Order)	
1	Laterites		a	Spodosols
2	Lateritic soils		b	Vertisols
3	Podzols		c	Alfisols
4	Black soils		d	Oxisols
5	Red soils		e	Ultisols

5.

A			B	
1	Aquic		a	Burnt
2	Udic		b	Dry
3	Ustic		c	Water
4	Xeric		d	Hot & Dry
5	Aridic		e	Humid

6.

A (Horizon)			B (Order)	
1	Spodic		a	Oxisol
2	Oxic		b	Inceptiols
3	Moltic		c	Aridisols
4	Cambic		d	Spodosol
5	Aridic		e	Mollisols

7.

	A			B
1	Ultisols		a	Ultimate
2	Mollisols		b	Pedalfers
3	Inceptisols		c	Dry
4	Alfisols		d	Soft
5	Aridisols		e	Begining

8.

	A (Order)			B (Soil type)
1	Alfisols		a	Laterite soils
2	Entisols		b	Red soils
3	Oxisols		c	Black soils
4	Vertisols		d	Desert soils
5	Aridisols		e	Alluvial soils

9.

	A (Mineral)			B (Property)
1	Iron minerals		a	Dull
2	Clay		b	Shining
3	Mica		c	Vitreous
4	Quartz		d	Metallic

10.

	A (Silicate clay)			B (Unique character)
1	Kaolinite		a	Brucite layer
2	Illite		b	Exchangeable cation & H_2O
3	Smectite		c	K^+
4	Chlorite		d	No interlayer

11.

	A (Term)			B (Defines)
1	Pedology		a	Study of rocks
2	Edaphology		b	Origin of rocks
3	Petrology		c	Description of rocks
4	Petrography		d	Study of soil from stand point of higher plants
5	Petrogenesis		e	Study of soil as natural body

12.

	A (Soil orders)			B (Soil forming process)
1	Histosols		a	Podzolization
2	Mollisols		b	Laterization
3	Spodosols		c	Paludization
4	Vertisols		d	Melanization
5	Oxisols		e	Pedoturbation

13.

	A (Mineral)			B (d - spacing ^{0}A)
1	Kaolinite		a	14
2	Smectite		b	28
3	Vermicultie		c	7.14
4	Glycerol Smectite		d	10 - 14
5	Swelled - Chlorite		e	17

14.

	A (Mineral)			B (CEC)
1	Kaolinite		a	10 - 40
2	Illite		b	80 - 120
3	Montmorillonite		c	150
4	Vermiculite		d	20 - 50
5	Allophane		e	3 - 15

15.

	A (Term)			B (Defines)
1	Dockuchaiev		a	The great soil groups of world
2	Glinka		b	First soil map
3	R.S.Murthy		c	Father of Pedology
4	Schokalskaya		d	Morpho-Genetic soil classification system
5	Marbut		e	Bench mask soils of India

TRUE OR FALSE

1. 3/4th of the earth crust is composed of earth metals

2. Density of core is more than crust and mantle

3. Quartz is a ferromagensian mineral

4. Tectosilicates are having 3 - dimensional frameworks arrangement of silicates

5. Haematite is a silicate mineral

6. Quartz is the light mineral

7. Metamorphic rocks formed from sedimentary rocks only through heat and pressure

8. Stratification is the most common feature of sedimentary rocks

9. Frost is effective agent of physical weathering than heat

10. The minerals lose their lustre and become soft on hydration

11. Fine grained feldspars decompose more rapidly than the coarse-grained feldspars when exposed to percolating water

12. Feldspar crystallize earlier than quartz and also weathers fastly

13. Most fertile soils in world are developed on alluvium

14. Southern exposed soils are warmer and subject to marked fluctuations in temperature and moisture

15. Well waters cooler in summer and warmer in winter

16. Silica: Sesquioxide ratio of tropical region will be >3

17. Laterization is removal of silica from upper layer

18. Parent materials having sufficient ferromagnesion minerals are conganial for development of laterites

19. Typical columnar structure is characteristic of solonetz

20. Churning is the process caused by swell-shrink clays

21. An organised listing of map units is called a series

22. An area which is completely circumscribed by a soil boundary on a map is called a soil delineation

23. Epipedons are synonyms with A horizon

24. Glossic horizon shows albic horizon characteristics gradually intruding into an argillic, a kandic or a natric horizon

25. Gypsic horizon at least 5 percent or more gypsum with 15 cm thickness

26. Accumulation of organic matter over the permafrost table is a character of cryoturbation

27. Slickensides are most common in Alfisols

28. COLE is helps to predict the potential of a soil to sell and shrink

29. Udic regime has low soil moisture than xeric

30. The diagnostic property of Histosols is having very high content of organic matter

31. Most of the vertisols develop on acidic parent material

32. Polished smooth ped surfaces of black cotton soils are called Slickensides

33. The best agricultural soils of world are mollisols

34. In aridisols evapotranspiration will be far less than rainfall

35. Pedalfers are non calcarious in nature

36. Alfisols are naturally fertile and un productive

37. Under acidic conditions, silica is mobile

38. Spodosls may also develop due to degradation of Alfisols

39. Oxisols are deeply weathered mineral soils that develop in basic parent materials

40. Though the clay content of oxisols are high, those clay is not sticky and plastic

41. *Perox* is suborder of oxisols

42. Soil crusting absent in oxisols

43. The major constraint of Histosol is high pore volume (low bulk density)

44. Volcanic ash materials are classified to Andisols

45. Andisols were earlier grouped under vertisols

46. Andisols are potentially very fertile when formed in basic volcanic ash

47. Expansion and contraction is very common in black soils of India

48. Desert soils are qualified for both aridisols and inceptisols

49. Red soils have relatively less base saturation than laterite soils

50. Salt affected soils are intrazonal soils

51. Saline soils are predominant in coastal regions

52. Wasteland is a synonym to degraded land

53. Color of a mineral may varies, but not streak

54. Fine powder of the mineral cannot represent its true color

55. AB horizon has characteristics of both an overlaying A- horizon and an overlaying B - horizon

56. AB horizon more like B than the A horizon

57. Presence of haematite indicates red color to soil

58. Ethylene glycol adsorption test recommended to determine total, internal and external surface areas of clay minerals

59. Muscovite is not a mica

60. X - ray diffraction technique is used for identification of minerals

61. Electron micrograph cannot be used for identification of minerals

62. Carbonation is a disintegration process

63. Sandy soils deposited by wind is loess

64. A smallest three dimensional portion is called a unit cell

65. Silica tertrahedron should have four oxygen ions surrounding one silicon

KEY TO CHOOSE THE CORRECT ANSWER

1. a	2. b	3. d	4. a	5. c	6. d
7. b	8. a	9. d	10. a	11. c	12. c
13. b	14. a	15. d	16. b	17. c	18. a
19. b	20. d	21. a	22. c	23. a	24. d
25. b	26. c	27. b	28. c	29. a	30. b
31. b	32. d	33. c	34. a	35. c	36. a
37. b	38. d	39. c	40. a	41. c	42. b
43. a	44. a	45. c	46. b	47. c	48. a
49. d	50. c	51. b	52. a	53. d	54. a
55. b	56. c	57. a	58. b	59. c	60. b
61. d	62. a	63. b	64. c	65. d	66. d
67. a	68. b	69. d	70. a	71. a	72. b
73. a	74. c	75. a	76. c	77. c	78. d
79. a	80. b	81. c	82. a	83. d	84. a

85. c	86. d	87. b	88. b	89. a	90. c
91. a	92. d	93. d	94. c	95. d	96. d
97. b	98. a	99. c	100. d	101. a	102. b
103. d	104. a	105. b	106. d	107. d	108. a
109. c	110. a	111. c	112. a	113. b	114. a
115. b	116. d	117. b	118. c	119. a	120. d
121. c	122. b	123. c	124. d	125. a	126. d
127. b	128. a	129. b	130. d	131. d	132. d
133. a	134. d	135. b	136. a	137. c	138. a
139. d	140. a	141. b	142. c	143. d	144. a
145. a	146. b	147. a	148. a	149. b	150. b

KEY TO FILL IN THE BLANKS

1. Dockuchaiev
2. Regolith
3. Weathering, Soil farming factors & process
4. 46.6%
5. $CaSO_4.2H_2O$
6. Secondary Minerals
7. One & Three
8. 2.85
9. Sedimentary rocks
10. intrusive or plutonic
11. Fine
12. acid
13. Pumice
14. Sedimentary
15. Weathering, transportation, deposition and diagenesis
16. Regolith
17. Reduction
18. Lessivage
19. Cool & Humid
20. Podzols / Spodosols
21. Warm & Humid
22. Solidization/ Dealkalization
23. 4-5
24. map-unit
25. Dockuchaiev
26. Sub group
27. epipedons
28. plaggen
29. Argillic horizon
30. Albic horizon
31. Durinodes
32. Fragipan
33. Aridic
34. 5

35. Central concept		36. Mollisol	
37. Ustic		38. ist	
39. Gelisols		40. Andisols	
41. Vertisol		42. Gilgai micro relief	
43. Vertisol		44. Calcification and Salinization	
45. Aridisols		46. Spodosols	47. Alfisols
48. Humid, warm		49. Ultisols	
50. Oxisols		51. Oxisols	
52. Black/ Vertisols		53. 3.0 to 3.5	
54. Gilgai		55. Desert	
56. Laterite		57. less than 2	
58. Global Assessment of soil degradation		59. Metallic	
60. Hardness		61. Cleavage	
62. Crystal form		63. Hue	
64. Value		65. Chroma	
66. Edaphology		67. Lattice Inter planar spacing of crystal	
68. Y - index		69. Detailed	
70. Pumice			

KEY TO MATCH THE FOLLOWING

1.	5-a	2-b	1-c	3-d	4-e
2.	3-a	5-b	1-c	2-d	4-e
3.	2-a	3-b	5-c	1-d	4-e
4.	3-a	4-b	5-c	1-d	2-e
5.	3-a	4-b	1-c	5-d	2-e
6.	2-a	4-b	5-c	1-d	3-e
7.	1-a	4-b	5-c	2-d	3-e
8.	3-a	1-b	4-c	5-d	2-e
9.	2-a	3-b	4-c	1-d	
10.	4-a	3-b	2-c	1-d	
11.	3-a	5-b	4-c	2-d	1-e

12.	3-a	5-b	1-c	2-d	4-e
13.	3-a	5-b	1-c	2-d	4-e
14.	2-a	3-b	4-c	5-d	1-e
15.	2-a	4-b	1-c	5-d	3-e

KEY TO TRUE OR FALSE

1. False [True: $3/4^{th}$ of earths crust is composed by O^{2-} & Si^{4+} which are non-metals]

2. True

3. False [True: It is a non - ferromagensian mineral]

4. True

5. False [True: It is a non - silicated mineral]

6. True

7. False [True: Metamorphic rocks formed from both igneous and sedimentary rocks]

8. True

9. True

10. True

11. True

12. True

13. True

14. True

15. True

16. False [True: It is < 2]

17. True

18. True

19. True

20. True

21. False [True: An organised listing of map units is called a legend]

22. True

23. False [True: Epipedons are not synonyms with A - horizon]

24. True

25. True

26. True

27. False [True: Slickensides are most common in vertisols]

28. True

29. False [True: Soil moisture is more in udic than xeric]

30. True

31. False [True: Vertisols develop on basic parent material]

32. True

33. True

34. False [True: In aridisols evapotranspiration exceeds the rainfall]

35. True

36. False [True: Alfisols are naturally fertile and productive]

37. False [True: Under acidic conditions, silica is immobile]

38. True

39. True

40. True

41. True

42. False [True: Soil crusting is one of major constraint in oxisols]

43. True

44. True

45. False [True: Andisols were earlier grouped under order inceptisol, suborder *Andepts*]

46. True

47. True

48. False [True: Desert soils are qualified for both aridisols and entisol]

49. False [True: Red soils have relatively high in base saturation status than laterite soils]

50. True

51. True

52. True

53. True

54. False [True: Fine powder of the mineral represents its true color]

55. True

56. False [True: AB horizon more like A than B horizon]

57. True

58. True

59. False [True: Muscovite & Biotite are micas]

60. True

61. False [True: Electron micrograph can use to identification of minerals having particle size less than 0.2 micron]

62. False [True: Carbonation is a decomposition process]

63. False [True: Sandy soils deposited by wind is Aeolian]

64. True

65. True

<table><tr><td>**2**</td><td># SOIL PHYSICS</td></tr></table>

P.N. Siva Prasad and Ramya S.H

FILL IN THE BLANKS

1. The relative proportion of soil particles is called as _______________

2. Soil texture considered as a __________ property of soil

3. _______________ is a permanent property of soil

4. Sand & silt fraction are called _______________

5. International pipette method also called as method _______________

6. Hydrometer method is rapid but less accurate than ___________ method

7. Mineral particles below __________ mm diameter are considered for mechanical analysis

8. Mineral particles having > 2mm in diameter are considered as __________

9. The physical property of the soil that does not change by cultural practices is _________

10. The physical property of the soil that is not affected by addition of organic matter in soil is ___________

11. The number of classifications in size groups soil particles by USDA system_________

12. The number of textural classes as per ISSS system is ________

13. In USDA system the number soil textural classes are _________

14. The lower size limit of sand in USDA classification of soil particles is ___________

15. According to the USDA system the diameter of silt particles is _________

16. According to the USDA system the diameter of the clay particles is ___________

17. The size limit of clay particles in ISSS system is ___________

18. Soil particles greater than 2 mm diameter is called ___________

19. In the field soil texture is generally bound by ___________ method

20. The method of finding different size groups of soil particles is known as __________

21. During mechanical analysis soil, organic matter is destroyed by treating soil with __________

22. Most commonly used dispersing agent in mechanical analysis of soil is __________

23. Finer soil particles are quantitatively determined by __________

24. The best agricultural soils are those that contain __________

25. Soils containing at least 70 % of sand and 15 % or less clay are said to be __________ soils

26. Soils containing equal proportion of sand, silt and clay particles are said to be __________

27. __________ refer to the resistance offered by the soil against the force that tends to deform it

28. What the value of particles density in stokes law __________

29. In stokes law η denotes __________

30. Textural analysis of all soils except calcareous and organic soils can be done by __________

31. More accurate analysis of soil texture can be done by __________

32. According to stokes law, the velocity of fall of a particle with the same density in viscous medium is directly proportional to __________ and inversely proportional to __________

33. Soils that have 60 % clay, 20 % sand and 20 % silt can be grouped into the textural class of __________

34. Arrangement of soil particles called as __________

35. Soil particles are separated based on their __________

36. __________ is the intermediate character between sand and clay

37. Soil having high amount of clay particles have __________ aeration property

38. Clay particles have __________ times higher specific surface area than same weight of sand

39. Macropores are high in __________ soils

40. Micropores are high in __________ soils

41. Difference in potential between fixed part and freely mobile part of diffused layer called as __________

42. Flesh of soil _______________

43. Fungi and bacteria produce Polysaccharides that favors soil_____________

44. _____________ clay is intermediate in its swelling and shrinkage characteristics

45. Best type of soil structure for favourable crop growth _________________

46. High amount of organic matter content in soil favours the formation of _____________ structure

47. _____________ type of structure restrict the free drainage

48. Spheroidal type of structure commonly found in ________________

49. Type of structure commonly found in sub surface of arid and semi arid _______________

50. Honey comb structural type found in _____________ soils

51. _________________ are the predominant cementing agents in laterite soils

52. Most prevalent method for testing the stability of soil aggregates _______________

53. When the soil particle arrangement is more in the horizontal axis than vertical axis then the soil structure is ____________________

54. Volumetric water content of soil is obtained by multiplying_____________ with bulk density of soil.

55. The dominant mineral in sand is _____________

56. _________________ and _____________ are called skeleton of soil

57. The forces acting on the spherical particles moving through a viscous medium are ___________

58. The components of fine earth are _______________

59. Surface area of spheroid can be calculated by which formulae _____________

60. Particles smaller then _________________ cannot be separated by sieving

61. Which is called as flesh of soil _____________

62. What is the depth of plough layer of soil ________________

63. What is the weight of a hectare of surface soil ________________

64. _______________ may be defined as arrangement of primary particles, secondary particles & voids in to certain definite pattern under field conditions

65. Soil structure described into ________________

66. What is the total number of soil structural classes ___________

67. When soil particles arrangement is more in the horizontal axis than vertical, then the structure is called ______________

68. In platy, when the units are thin, then soil structure is called as _____________ structure

69. In platy, when the units are thick then soil structure is called as _____________ structure

70. Vertical axis is more developed than other soil structure in ___________ structure

71. In prism structure, when the top of prisms are still plane, level and clean cut called as _______________

72. In prism structure, when the top of such ped is rounded is called as

____

73. All the three dimensions are about the same size and the peds are cube like with flat or rounded faces is called as ______________

74. In blocky structure of the soil, when the faces and edges are mainly rounded is called as _________________

75. When the granules are spheroidal and porous is called as

____

76. When the granules are spheroidal and non-porous is called as

____

77. Classification of soil structure for field description is determined based on ___________ shape & arrangement of peds

78. ___________ Structure promote good drainage, aeration and root penetration

79. The manifestation of physical forces of adhesion and cohesion at various moisture contents with in the soil is called ______________

80. Atterburg constant is related to the _________________ soil property

81. The optimum consistency for the tillage operation to produce aggregates of suitable size is ___________

82. The difference between upper plastic and lower plastic limits of soil consistency _____________

83. The attractive force between soil solid particle is ____________

84. The force with which water is held to the soil particle is called

____

85. The water content at which the volume of soil remains constant regardless of drying (particularly in clay soil) is called ______________

86. Soil crust is formed due to __________________

87. Surface crusting is a major problem in __________________ soils of Andhra Pradesh

88. The soil texture highly susceptible to crusting ______________

89. Plasticity and stickness is the characteristic feature of _____________

90. Fragments are ______________

91. Naturally formed soil aggregates are called as _____________

92. Natural aggregate which vary in their water stability is called __________

93. Poorly formed, nondurable, indistinct peds &broken into many broken peds are _________

94. Moderately well developed peds fairly durable distinct are __________

95. Very well formed fairly durable distinct are ____________

96. The bonding of the soil particles into structural units is called _______________

97. _____________ is used for coherent mass of soil broken into any shape by artificial means such as tillage

98. Broken peds are known as ________

99. _________ is a coherent mass of soil formed with in soil by precipitation of certain chemicals dissolved in percolating water

100. The shape and arrangements of peds is designated as __________

101. Classification of soil structure for field description is determined based on _____________ by size of peds

102. Natural cleavage plains are distinct peds in __________________

103. When spheroidal peds are porous, the structure is called as ______________

104. Less porous peds are called _____________

105. Edges are sharp & rectangular faces distinct they are called ____________

106. In block like structures edges are round faces distinct they are called _______________

107. Granular pores are _________

108. _________ are <0.5c.m.

109. In soil structure classification, the term type refers to the ___________and_______________of the aggregates

110. In soil structure classification, the term class refers to _____________ of the aggregates

111. Classification of soil structure for field description is determined based on _________ by shape & distinctness of durability of peds

112. The structure classification according to durability of peds is known as_________

113. Natural cleavage plains are absent or indistinct in ______________

114. Single grain structure occurs in ____________

115. Massive structure is coherent mass with high bulk density, occurs in ______________soils

116. Massive or loose aggregates of nodular ferruginous mass is ______________________

117. Vesicular or honeycomb structure can be seen in ______________

118. Block like structures found in ____________

119. In block like structures compaction of clayey soils with heavy machinery ______________

120. In soil structure classification, the term grade refers to the __________ of the aggregates

121. The prismatic structure is found in sub soils rich in ____________

122. Natural cleavage plains are distinct in _____________ structure

123. No cleavage lines in puddle soils form a _____________ field

124. Prominent structure in black soils is _________________

125. The first step in the formation of stable aggregates is _______________

126. The two steps in the formation of stable aggregates are __________ and _____________

127. Which one of the following is tendency to peptize clay __________

128. Element essential for the flocculation of soil particles is _________

129. Element essential for the deflocculating of soil particles is ________

130. Zeta potential of soil particles reduced by ________

131. Increase in zeta potential of soil particles leads to ______________

132. Structural management of coarse textured neutral soil is done by application of_____________

133. Synthetic chemicals added to the soil to improve the soil structure are called_________

134. The degraded structure of alkali soil is restored by the application of _____________

135. Soil aggregation can be improved by the addition of_____________ fertilizers

136. Specific surfaces area is highest for _____________ soil

137. Crops which help in building soil structure are _____________

138. The most favored soil structure desired for agriculture is _____________

139. Theory of bonding of clay-clay particles was given by _____________

140. Anion responsible for soil aggregation is _____________

141. Columnar structure is mostly found in _____________

142. The physical property will not change on compaction is _____________

143. Volume of soil not occupied by soil solids is referred to as _____________

144. Due to compaction, the particle density of soil _____________

145. The particles density value of mineral soil is generally taken as _____________

146. Compaction of soil increases _____________

147. Bulk density also called as _____________

148. Particle density also called as _____________

149. The bulk density of a particular soil increases its pore space _____________

150. Increase in bulk density of soil by dynamic loading is called _____________

151. The strength of soil crust is measured with the help of _____________

152. Sub soils high in _____________

153. _____________ is associated with deterioration of soil structure

154. Thickness of crust increases with increases in _____________ content

155. Montmorillanite clay forms thick _____________

156. _____________ clay forms thin crust

157. _____________ can be evaluated using penetrometer, ballon pressure technique, modules of rupture test

158. Modulus of rupture is a _____________

159. _____________ is a dynamic physical property of the soil

160. Property of toughness and capacity to be molded is called _______________

161. Water content at which soils practically liquid but possess a small shearing strength. This is termed as _______________

162. Any soil can exhibit plasticity it contains more than _____________ Clay

163. Higher the hydration energy of adsorbed cation _______________________

164. Addition of organic matter decreases the _________________

165. Atterberg's limit is related to the _____________ soil property

166. Soil moisture condition at friable consistency is ___________

167. Plasticity is highest for ___________ soils

168. Tilling fine textured soils is difficult due to their _______________

169. Higher the clay content in the soil offers more surface area thereby _______________

170. Shrinkage limit is attained at that water content at which the volume of the soil remains ___________

171. The expansiveness of soil can be quantified as _________________

172. COLE value is <0.03 it is _______________

173. COLE value is > 0.09 it is___________

174. COLE = ___________________________

175. In soil the volume occupied by water and air together is called _______________

176. The ratio of total volume of pore spaces to the total volume of soil is called as ___________

177. The ratio of total volume of air spaces to the total volume of soil is called as ____________

178. _________________ is the index of relative air content of the soil

179. The ratio of total volume of pores to the total volume of soil solids is called _______________

180. _____________________ is an index of the relative volume of soil pores

181. The ratio of total volume of water occupied in the pore spaces at a specified time to the total volume of soil is called as _________________

182. The ratio of volume of water present in the soil at a particularly time to the volume of pores is called as _________________

183. Strength of the soil positively related to it's ________________

184. Total pore spaces for unit volume of soil is more in ________________

185. Soil micropores having diameter of ________________

186. With increase in depth of soil, the bulk density ________________

187. With an increase in organic matter content of the soil, there is decreases in ________________

188. Compaction of soil results in increases of its ________________

189. If bulk density and particles density of a soil are 1.19 g/cc and 2.65 g/cc respectively than the per cent pores space will be ________________

190. If 100 Cm^3 of soil has an oven dry weight of 135 grams, what is its bulk density ________________

191. If 100 Cm^3 of soils is saturated contains 40 g of water, what is the per cent spaces ________________

192. Bulk density of fine textured soils is ________________

193. Bulk density of coarse textured soils is ________________

194. Bulk density of organic peat soils is ________________

195. The ratio of total volume of soil to the total mass of dry soil is called as ________________

196. Bulk density in soils can be estimated by ________________

197. A characteristic red colour is important to the soils by the ________________

198. In muncell colour notation the purity of colour is indication by ________________

199. In muncell colour notation the dominant spectral colour is indicated by ________________

200. In muncell colour chart, the brightness of colour is denoted by the term ________________

201. What is the diameter of water molecule ________________

202. What is the angle of hydrogen atoms from oxygen atoms in water molecules ________________

203. What is the amount of energy released when water freezes into ice ________________

204. The value of atmospheric pressure is ________________

205. One Mega Pascal (MPa) ________________ bars

206. One bar equivalent to _______________ Mega Pascal

207. One bar equivalent to __________ kilo Pascal's

208. One bar is equal to _____________ height of water column

209. One bar is equal to _________________

210. 1 bar or 101.3 k Pa is equal to pressure exerted by _________________

211. The density of water is maximum at __________

212. The free energy of pure water is always ____________

213. The soil water potential in saturated soil is ___________

214. The soil water potential at air dryness is ________________

215. Soil water tension or potential at field capacity ________________

216. Soil water tension at permanent wiliting point is ______________

217. What is the suction of partially available water in soil __________

218. What is the suction of available water in soil ________

219. Soil moisture tension at hygroscopic coefficient is _________________

220. The soil water potential at air dryness is ________

221. Tensiometer measure soil moisture tension upto _________

222. Soil water potential of -10 J/kg is equal to soil water suction of __________

223. What is the dissociation constant of water ________________

224. Ionic product of water is ________________

225. Soil water potential always has ________________ sign

226. Soil water suction always has ______________ sign

227. In water molecule, the angle formed between two atoms of hydrogen with oxygen is ________________ degrees

228. The process of water entry into soil through the surface may be either downward or lateral or both is called as _________________

229. The most important factor influences the infiltration of water is _________________

230. In which soils water infiltration is more rapid _____________

231. As per Darcy's law , the flux equals to hydraulic conductivity when hydraulic gradient becomes ______________

232. As per Darcy law, flux of water is proportional to hydraulic gradient

233. According to Poiseuilles law, the rate of flow of a liquid through a narrow tube is directly proportional to the ______________ of the radius of the particle

234. Removal of excess water from soil is known as ______________

235. The relation between rise of water (h) and diameter of a capillarity (d) both in centimeters is h= ______________

236. When salts are present in soil water its free energy ______________

237. The retention of water in the soil is due to ______________ and ______________ forces of capillarity

238. The capillary rise of water is due to the forces of ______________

239. The component of total soil water potential due to dissolved salts ______________

240. Saturated conductivity of water is highest in ______________ soil

241. The rate of saturated flow water in soil is in the order of ______________

242. The rate of unsaturated flow of water in soil is in the order of ______________

243. The concept of pF was given by ______________

244. As the water content of soil increases the soil moisture tension ______________

245. In saturated soils the rate of flow of liquid varies with 4th power of ______________ of the pores

246. Rate of flow of liquid through a porous medium under unit hydraulic gradient ($K = Q/i$) is called as ______________

247. Total hydraulic head per unit distance in the soil water column $\left[\dfrac{\Delta H}{L}\right]$ is called as ______________

248. Hydraulic conductivity of soil is found to increase in which order ______________

249. The potential results from adsorption and capillary forces of the soil matrix acting on soil water is called as ______________

250. The potential developed due to the attraction of water molecules by soil solids is called as ______________

251. The potential result from earth gravitational force is called as ______________

252. Osmotic potential is a combination of ______________

253. Water retained between field capacity and permanent wilting point (33-1500 kPa) is known as _______________

254. Water held in the pores and drained by 100 cm of suction is called as _______________

255. Water molecule is __________ in shape

256. At normal temperature is a liquid and not a gas because of ……..bonding

257. Kinetic energy of water is minimum in ________ form, more in ________ form and maximum in ________ form

258. The height of raise of water in a capillary tube is __________ proportional to the surface tension of water

259. The rate of capillary raise of water is maximum in __________ and minimum in __________ soils

260. Evaporation of water results in ____________ effect

261. Attraction of water molecules for each other is known as_______________

262. The tenacity with which water molecules is held by soil matrix is termed as __________________

263. The amount of water held in soil after excess water has drained away is called as _________________

264. In ice each water molecules is linked to ___________ water molecules

265. One cubic centimeter of liquid water contain _______________ number of water molecules

266. Water exhibit maximum density at ________

267. Below_________ ^{0}C the volume of water increases

268. At 25 ^{0} C the ionic product of water ________

269. Capillarity is mainly due to _______________and ____________force

270. The water holding capacity of sandy soils increased with addition of _______________

271. Available water held between ________ and ______bars

272. Upper limit of available water _________________

273. Water held by surface tension in the form of continuous film around soil particles is called as _________________

274. In gravimetric method of moisture estimation wet soil is dried in an oven at a temperature of _______________

275. Boundary layer between air and water _______________

276. Meniscus formed in a capillary tube is _______________ form

277. The phenomenon of capillarity is due to _______________

278. In loamy soils moisture equivalent is almost equal to _______________

279. In heavy soils the moisture equivalent is _________ than field capacity

280. Gypsum block function effectively over range of _________ bar

281. Water held against the force of gravity in micropores is _______________

282. Capillary water in the soil is that which is found in the _______________

283. Water held in thin films around soil particles is _______________

284. Hygroscopic water can be removed from soil by _______________

285. The inner film of water around soil particles is held by _______________

286. When soil is saturated the dominant force affecting water movement is _______________

287. Percentage of pores which is drained by 100 cm of suction is called as _______________

288. Soil texture in the order of decreasing water retention capacity _______________

289. Soils in the order of decreasing water retention capacity _______________

290. The energy released when ions become hydrated is called as _______________

291. When the clay particles are hydrated, the energy released is called as _______________

292. In a saturated soil , the pressure potential is _______________

293. In water unsaturated soil, the pressure potential is _______________

294. Total soil water potential t= _______________

295. The amount of energy required to evaporate one gram of water _________

296. 10 cm of water column is equal to the pF scale of _________

297. What is the value of pF scale at field capacity _________

298. What is the value of pF scale at oven dry condition _________

299. What is the value of pF scale at wilting point _________

300. What is the value of pF scale at hygroscopic point _________

301. In the field matric potential of soil is measured by _________________

302. The instruments used for the measured of negative soil water tension _______________

303. Tensiometer can be used to measure soil water pressure upto _______________

304. Tensiometer can be used to measure soil water tension or suction upto ____________

305. Soil water potential in the field can be measured by _________________

306. Soil water potential in the laboratory can be measured by _________________________________

307. In the laboratory moisture content can be measured by porous cup apparatus upto ________

308. In the laboratory moisture content can be measured by pressure plate apparatus upto ____________

309. Non- destructive method of measuring soil water content is _______________

310. Insitu measurement of soil water can be done by ________________________

311. Which one of the materials used in electrical resistance method of soil water estimation ____________

312. What is the source of gamma radiation in gamma ray scanner used for soil water estimation ____________

313. In neutron probe the slow neutron are absorbed by ________ gas

314. Pressure membrane operates detector is used for the determination of soil _________________

315. The steady infiltration is approximately equal to ____________ at saturation

316. Under saturated conditions flow rate of water in soils is proportional to _______________

317. The rate of unsaturated flow water in soil is in the order of _________________________

318. Under unsaturated conditions the driving force for water movement in soils is _______________

319. In unsaturated soil the, the macropores filled with ________

320. Under unsaturated conditions the driving force for water movement in soils is _________________________

321. The attraction of water from the atmosphere by the soil solids is called

____

322. The rate at which the conductor warms up uniformly is known as

____

323. Thermal conductivity is maximum for _______________

324. The fraction of incident radiation that is reflected by the land surface is termed as _________

325. Ratio of heat supplied to a body to the corresponding rise in temperature is known as _______________

326. Quality of the heat passing in a unit time through a unit area is known as

____

327. Thermal conductivity of soil varies in the order _______________

328. The ratio of thermal conductivity to heat capacity is known as

____

329. The ability of the body to retain heat is _______________

330. The amount of heat existing in the body is called _________

331. The flow of heat through an unequally heated body from the high temperature to low temperature is known as _______________

332. Transparent plastic mulch leads to _______________

333. The reciprocal of thermal diffusivity is _______________

334. Low soil temperature results in _______________

335. The specific heat of soil forming mineral is about _______________

336. Specific heat of water is about 0.5 times more than that of

____

337. Thermal conductivity of different soils is in the order _______________

338. Heat transfer in the soil solids takes place mainly by _______________ process

339. The main source of heat energy for the soil is _________

340. The microbial activity is restricted at temperatures below _________

341. _______________ is the ratio of heat capacity to thermal conductivity

342. The concentration of CO_2 in earth crust is _________

343. The _________ content of soil air is more than that in atmospheric air

344. Movement of gas is due to concentration gradient is called as

____

345. The major mechanism (more than 90 %) of gaseous exchange through soil is ____________

346. The process of heat transfer from one body to another in the form of electromagnetic waves is called ________________

347. For normal crop growth, aeration porosity of soil must be not less than __________

348. ODR value optimum for most of the crops is ________________________

349. Drainage improves the ____________________

350. Which one of the following in soil air helps to dissolve the nutrients in soil ____________

351. The author of the reference book "Fundamental of soil physics" is ________________

CHOOSE THE CORRECT ANSWER

1. The size of boulders is
   (a) 2.0 – 75mm   (b) 250 - 600mm   (c) >600 mm   (d) 2 – 75 mm

2. Dispersion is carried out by means of chemical dispersing agent called
   (a) Sodium hexametaphosphate   (b) Sodium bicarbonate
   (c) Dil. Hcl   (d) Hydrogen peroxide

3. Viscosity is directly proportional to the square of
   (a) Velocity   (b) Density of solid
   (c) Radius   (d) Density of fluid

4. According to STOKE's law particle size must be greater
   (a) < 0.1 mm   (b) 0.001 mm
   (c) > 0.001 mm   (d) 0.2 mm

5. Which of the following are deflocculating agents
   (a) Sodium oxalate   (b) Sodium pyrophosphate
   (c) Sodium phosphate   (d) All the above

6. Number of classes in USDA classification
   (a) 12   (b) 8   (c) 9   (d) 10

7. The particle size of the fine earth
   (a) 0.002 mm   (b) < 2 mm   (c) < 02 mm   (d) < 0.02 mm

8. The group of mineral particles of soil separated based on size they are referred as

   (a) Soil separates   (b) Soil group

   (c) Soil particle   (d) Soil classification

9. The particle size of soil particle

   (a) 0.002mm   (b) < 2 mm   (c) < 02 mm   (d) < 0.02 mm

10. Which of the following institution not classified the soil separates

    (a) ISSS (international society of soil science)

    (b) USDA (united states department of agriculture)

    (c) BSI (British standard institution)

    (d) USSC (united states soil conservation service)

11. Size of clay particles according to USDA

    (a) 2 - 0.05mm   (b) 2 – 0.2 mm

    (c) 2 – 0.02 mm   (d) 2 - 0.002 mm

12. Largest soil particle

    (a) Sand   (b) Stone   (c) Pebble   (d) Gravel

13. The soil separate which is resistant to weathering

    (a) Sand   (b) Clay   (c) Silt   (d) All

14. Dominant mineral found in sand

    (a) Quartz   (b) Feldspar   (c) Mica   (d) None

15. Stony soils are _________ than normal soil

    (a) Warmer   (b) Cooler   (c) Both   (d) None

16. Sandy soils are

    (a) Non sticky   (b) Non plastic

    (c) Non cohesive   (d) All

17. Particle size of silt fraction

    (a) 0.02 – 0.002 mm   (b) < 2 mm

    (c) 0.2 – 2 mm   (d) < 0.002mm

18. Smallest soil particles

    (a) Sand   (b) Silt   (c) Clay   (d) Pebble

19. Clay particles are ___________ type of structure

    (a) Platy   (b) Blocky   (c) Spheroid   (d) Crumb

20. Most active part of soil

    (a) Sand   (b) Silt   (c) Clay   (d) Pebble

21. Chemically inactive part of soil
    (a) Clay   (b) Sand   (c) Silt   (d) Gravel

22. The following soil separate exhibit the Brownian movement
    (a) Sand   (b) Clay   (c) Silt   (d) Gravel

23. Clay soil have which of the following property
    (a) Low water holding capacity   (b) High macro pores
    (c) High buffering capacity   (d) Low porosity

24. Soil textural classes indicates
    (a) Physical property   (b) Chemical property
    (c) Both   (d) None

25. How many fundamental group of soil textural classes are there
    (a) 3   (b) 4   (c) 8   (d) 7

26. Among the following which is the fundamental group of soil texture
    (a) Sandy loam   (b) Sandy clay loam (c) Loam   (d) All

27. Equal proportion of sand silt and clay called as
    (a) Marl   (b) Loam   (c) Both   (d) All

28. Following soil texture is good for cultivation
    (a) Sandy   (b) Clay   (c) Loam   (d) Silty soils

29. Clay soils contain atleast ______________ % clay
    (a) 35 %   (b) 50%   (c) 20 %   (d) 47 %

30. Total porosity high in ___________ soils
    (a) Sandy   (b) Clay   (c) Silty   (d) Loamy

31. Internal drainage is good in __________ soils
    (a) Clay   (b) Sand   (c) Loam   (d) All

32. Silty soils contain atleast ________ % silt
    (a) 80   (b) 90   (c) 65   (d) 55

33. Light and heavy soils are classified based on
    (a) Size   (b) Structure
    (c) Weight   (d) Ease of cultivation

34. Feel method used to determine
    (a) Structure   (b) Texture
    (c) Particle size   (d) Porosity

35. Sieve used for preparation of soil for analysis
    (a) 2mm   (b) 1mm   (c) 0.2 mm   (d) 20mm

36. Dispersing agent used in mechanical analysis
    (a) $CaCO_3$
    (b) NaOH
    (c) Sodium hexametaphosphate
    (d) Both b & c

37. Chemical used in destruction of carbonates and exchangeable metals
    (a) Dilute HCl   (b) NaOH   (c) $CaCO_3$ (d) NaCl

38. Fractionation method is used in separation of __________ fraction during mechanical analysis
    (a) Sand   (b) Silt   (c) Clay   (d) Both a & b

39. Sedimentation method used in separation of ___________ fraction during mechanical analysis
    (a) Sand   (b) Silt   (c) Clay   (d) Both a & b

40. Who developed the international pipette method for mechanical analysis
    (a) G.W. Robinson
    (b) Atterberg
    (c) Andreson
    (d) Bouyoucos. G

41. Mechanicl analysis is based on ___________
    (a) Poisuleslaw   (b) Stokes law   (c) Darcy law   (d) Ficks law

42. Base side of the triangular textural diagram represented
    (a) % sand   (b) % clay   (c) % silt   (d) none

43. Flocculation power of cation increases with ___________
    (a) Increasing valency
    (b) Decreasing electron
    (c) Increasing electron
    (d) Both a & b

44. Correct order of dispersion power of cation
    (a) Monovalent> divalent > trivalent
    (b) Divalent > monovalent > trivalent
    (c) Trivalent< divalent < monovalent
    (d) Both a & c

45. pH of cat clay soils
    (a) < 3.5   (b) > 3.5   (c) <5   (d) < 6.5

46. Acid sulphate soils are formed due to
    (a) Reduction   (b) Oxidation   (c) Dispersion   (d) None

47. The organism responsible for development of acidity in acid sulphate soils
    (a) *Thiobacillus ferroxidants*
    (b) *Dislphovibrio*
    (c) *Pseudomonas fluroscence*
    (d) *Bacillus subtilis*

48. Acid sulphate soils can managed by
    (a) Control water table　　　　　　(b) Keeping area flood
    (c) By growing paddy　　　　　　　(d) All

49. Soil texture contain more sand among the following is
    (a) Sandy loam　　　　　　　　　　(b) Silty loam
    (c) Silty clay loam　　　　　　　　(d) Loamy sand

50. Specific surface area of montmorillonite in the range of __________ m^2/g
    (a) 100 -300　　(b) 300 – 500　　(c) 500 -800　　(d) 800 – 1000

51. Soil aggregates are
    (a) Primary particles　　　　　　　(b) Secondery particles
    (c) Tertiary particles　　　　　　　(d) None

52. Among the following which on is artificially formed
    (a) Ped　　　　　　(b) clod　　　　　　(c) fragments　　(d) all

53. Broken peds are called as
    (a) Ped　　　　　　(b) Clod　　　　　　(c) Fragments　　(d) Pedon

54. The resultants formed by precipitation and consolidation of chemicals dissolved in percolation water
    (a) Concretion　　(b) Fragments　　(c) Ped　　　　(d) Clod

55. Ability of soil to restore its structural form through natural process when stress arose from process like tillage traffic and rain drop are reduced or removed
    (a) Structure vulnerability　　　　(b) Structural resiliency
    (c) Structural stability　　　　　　(d) Structure form

56. Type of soil structure classified based on
    (a) Size　　　　　(b) Shape　　　　(c) Stability　　(d) All

57. Loamy soils considered as best for crop production because
    (a) They retain more water and nutrient than sand
    (b) They have better drainage and aeration than sand
    (c) They have better tillage property than clay
    (d) All

58. The type soil structure where in horizontal axis is longer than vertical axis
    (a) Platy　　　　(b) Prism　　　　(c) Blocky　　(d) Spheroid

59. Most suitable type of soil structure for agriculture
    (a) Crumb　　　　(b) Granular　　(c) Platy　　(d) Blocky

60. Among the following soil structure which one commonly found on sub soils

    (a) Granular   (b) Crumb   (c) Blocky   (d) Spheroid

61. Classes of soil structure classified based on

    (a) Size   (b) Shape   (c) Both   (d) None

62. ___________________ is a dynamic property of soils that varies with the variation of soil moisture and applied stress.

    (a) Soil texture   (b) Soil structure

    (c) Soil pH   (d) Soil consistency

63. Soil compaction increases the

    (a) Particle density   (b) Bulk density

    (c) Porosity   (d) Permeability

64. The distinctness and durability of peds is used to know

    (a) Grade of the soil   (b) Type of the soil

    (c) Class of the soil   (d) None

65. Soil compaction decreases the

    (a) Particle density   (b) Hydraulic conductivity

    (c) Porosity   (d) Permeability

66. Soil crust strength can be evaluated by using

    (a) Penetrometer   (b) Balloon pressure technique

    (c) Modulus of rupture test   (d) All

67. Which of the following are influenced by soil structure

    (a) Nature of the porosity of soils   (b) Water holding capacity

    (c) None   (d) Both a & b

68. Among the following which is /are mainly responsible for cementation of primary particles into stable

    (a) Clay particles   (b) Hydrous oxides of iron and Al
    (c) Organic substances   (d) All

69. The following primary soil particle/s play key role in soil aggregation

    (a) Sand   (b) Silt   (c) Clay   (d) All

70. Soil dominated by kaolinite clay produce ___________ structure

    (a) Platy   (b) Blocky   (c) a and b   (d) None

71. Correct order of cations with respect to the ability to flocculate clays

    (a) $Ca^{2+}>K^+>Na^+>Al^{3+}$ (b) $Al^{3+}>Ca^{2+}>K^+>Na^+$

    (c) $Al^{3+}>Ca^{2+}>Na>K^+$ (d) $Al^{3+}<Ca^{2+}<Na^+>K^+$

72. Among the microorganisms which of the following have more direct effect on aggregate formation
    (a) Fungi   (b) Bacteria
    (c) Actinomycets   (d) None

73. Factors which is crucial for providing structural stability
    (a) Air   (b) water
    (c) Mineral content   (d) All

74. Application of following fertilizer improves the soil structure
    (a) SSP   (b) Urea   (c) $Na(NO_3)_2$ (d) MOP

75. Application of following improves the soil structure
    (a) Lime stone   (b) Dolomite
    (c) Green manure   (d) All

76. In alkali soils following chemicals are used to improve soil structure
    (a) Gypsum   (b) Pyrite   (c) H_2SO_4 (d) a and b

77. Type of soil which are most susceptible to formation of crust
    (a) Clay loam   (b) Sandy
    (c) Silty clay loam   (d) Loam

78. Strong crust have
    (a) Low BD   (b) high BD   (c) High WHC (d) None

79. Puddled paddy field usually have following type of soil structure
    (a) Massive type   (b) Crumby   (c) Granular   (d) No structure

80. Prism like structure are mainly seen in
    (a) Alkali soils   (b) Saline soils
    (c) Alluvial soils   (d) Laterite soils

81. Cementing agent responsible for aggregation in laterite soils
    (a) Organic matter   (b) Calcium carbonates
    (c) Hydrous oxides of iron and Al (d) Clay

82. Among the following the most important cementing agent in the formation of aggregation
    (a) Plant nutrient   (b) Salts
    (c) Microorganisms   (d) Humus

83. Alkali soils are reclaimed by application of
    (a) Organic matter   (b) Gypsum
    (c) Dolomite   (d) Lime stone

84. Coarse texture soils can be managed by application of
    (a) Soil amendments   (b) Organic matter
    (c) Fertilisers   (d) Mulching

85. Application of which of the following reduce the soil crusting
    (a) Fertiliser   (b) Amendment
    (c) FYM   (d) Surface mulch

86. Very poor soil structure is found in
    (a) Saline soils   (b) Acid soils
    (c) Sodic soils   (d) Saline sodic

87. Among the following highly porous structure is
    (a) Platy   (b) Blocky   (c) Granular   (d) Crumby

88. Crops most suitable for formation of best type of soil structure
    (a) Maize   (b) Mustard   (c) Rice   (d) Grasses

89. Direct method of soil structure evaluation is
    (a) Microscopic method   (b) Macroscopic method
    (c) Wet sieving   (d) Pipette method

90. Carbon dioxide content of soil air will be highest in
    (a) Well aggregated soil   (b) Bare soil
    (c) Un manured soil   (d) Manured soil

91. Nitrogen content of soil air is
    (a) More than   (b) Less than
    (c) Almost same as atmospheric air   (d) None of these

92. The normal or standard atmospheric pressure is equal to
    (a) 1013 mill bar   (b) 76 mill bar
    (c) 980 mill bar   (d) None of these

93. Well aerated soil air contains how many times more carbon dioxide than
    atmospheric air
    (a) 8 to 10   (b) 20 to 100   (c) 10 to 20   (d) 100 to 300

94. Diffusion of gases in between soil air and atmospheric air takes place
    following
    (a) Boyle's law   (b) Charles law   (c) Fick's law   (d) Fourier's law

95. The most important mechanism for renewal of soil air is
    (a) Mass flow   (b) Diffusion   (c) Autolysis   (d) Osmosis

96. The different components of soil air is separated by using
    (a) Graham's law
    (b) Fick's law
    (c) Boyle's law
    (d) Charles' law

97. In an anaerobic environment which one of the following can act as electron acceptor (*i.e.*oxygen)
    (a) Oxygen
    (b) Hydrogen
    (c) Nitrate
    (d) None of them

98. In soil, the normal rate of carbon dioxide production is about
    (a) 2 litres/m^2/day
    (b) 7 litres/m^2/day
    (c) 10 litres/m^2/day
    (d) 15 litres/m^2/day

99. For normal growth of soil organisms the ODR values should be at least
    (a) 10×10^{-8} gm/cm^2/min
    (b) 40×10^{-8} gm/cm^2/min
    (c) 50×10^{-8} gm/cm^2/min
    (d) 100×10^{-8} gm/cm^2/min

100. The percentage of carbon dioxide in atmospheric air is
    (a) 0.003
    (b) 0.03
    (c) 0.3
    (d) 0.25

101. If an alluvial forest soil is in flooded condition for one week, anaerobic organisms become active and the gases other than nitrogen , carbon dioxide and oxygen are expected evolve is
    (a) H_2S, CH_4 and N_2
    (b) H_2S, CH_4 and N_2O
    (c) H_2S, CH_4 and C_2H_4
    (d) $H_2S, CH_4 , C_2H_4 , N_2O$ and N_2

102. Platinum electrode along with a reference electrode is generally used to measure
    (a) EC
    (b) Eh
    (c) pH
    (d) pF

103. Root growth of most plants ceases due to lack of oxygen supply when ODR in gm/cm^2/min is
    (a) 10×10^{-8}
    (b) 30×10^{-8}
    (c) 20×10^{-8}
    (d) 40×10^{-8}

104. For satisfactory growth of most plants and microorganisms the ODR in gm/cm^2/min should be at least
    (a) 20×10^{-8}
    (b) 30×10^{-8}
    (c) 40×10^{-8}
    (d) 50×10^{-8}

105. Aeration porosity is
    (a) Total porosity- volumetric water content
    (b) Total volume of soil-total volume of soil solid
    (c) Total volume of soil-total porosity
    (d) Total porosity- gravimetric water content

106. Graham's law of diffusion is used
    (a) To know the rate of diffusion of gases
    (b) To known the mass flow of gases
    (c) To separate the different components of gaseous mixture
    (d) To know the renewal of soil air

107. Radiation from a very hot body, e.g the sun,is in
    (a) Short waves   (b) Medium waves
    (c) Long waves   (d) Ultra waves

108. Radiation from heated soil is in
    (a) Long waves   (b) Short waves
    (c) Medium waves   (d) Ultra waves

109. Evaporation is ________________ process
    (a) An exothermic   (b) An isothermic
    (c) An endothermic   (d) A mesothermic

110. Condensation is ____________ process
    (a) An exothermic   (b) An isothermic
    (c) An endothermic   (d) A mesothermic

111. Specific heat of dry mineral soil is
    (a) More than of moist soil   (b) Equal to that of moist soil
    (c) Less than of moist soil   (d) None

112. The transfer of heat through the emission of energy in the form of electro-magnetic waves from bodies above ^0K is referred to
    (a) Conduction   (b) Radiation
    (c) Convection   (d) None of these

113. The propagation of heat within a body by internal molecular motion is known as
    (a) Conduction   (b) Radiation   (c) Convection (d) None of these

114. The transfer of heat energy through the movement of a heat carrying mass is called
    (a) Conduction   (b) radiation   (c) Convection (d) None of these

115. Which soil has the highest thermal conductivity
    (a) Peat   (b) Silt loam   (c) Clay loam   (d) Loamy sand

116. Which law follows the heat flow through soil
    (a) Darcy's law   (b) Fourier's law   (c) Ohms law   (d) Fick's law

117. Increase in soil temperature
    (a) Increases hydraulic temperature
    (b) Decreases hydraulic conductivity
    (c) Causes no changes in hydraulic conductivity
    (d) None of these

118. Which soil has lowest thermal conductivity
    (a) Sand   (b) Clay   (c) Loam   (d) Peat

119. The most sensitive type of resistance thermometer is
    (a) Thermocouple thermometer   (b) Mercury thermometer
    (c) Bimetalic thermometer   (d) Thermistor thermometer

120. Which thermometer works on the principle of expansion of solid
    (a) Mercury thermometer   (b) Bimetalic thermometer
    (c) Thermocouple thermometer   (d) Thermister thermometer

121. Which of the following would be the most important factor in influencing
    soil temperature
    (a) Humus content   (b) Ca content
    (c) Fe content   (d) Water content

122. Air temperature can be recovered continuously with
    (a) Barograph   (b) Hydrograph
    (c) Thermograph   (d) None of these

123. Albedo is the
    (a) Thermal coefficient of the surface towards short wave radiation
    (b) Refractivity   (c) Reflectivity   (d) Incidental

124. The heat that radiates back from earth is known as
    (a) Solar radiation   (b) Insulation
    (c) Terrestrial radiation   (d) None of these

125. Solar radiation having more energy potential has
    (a) Short wave length   (b) long wave length
    (c) Medium wave length   (d) Ultra wave length

126. Specific heat of liquid water as compared to that of ice is
    (a) More   (b) Equal   (c) Less   (d) None of these

127. Latent heat of vaporization of water at boiling point is
    (a) 1 cal/gm   (b) 80 cal/gm   (c) 537 cal/gm   (d) 590 cal/gm

128. The amount of heat energy required to evaporate a layer of 1 cm thick of water covering 1 hectare of land is

    (a) 40×10^6 cal   (b) 54×10^6 cal   (c) 58×10^6 cal (d) 80×10^6 cal

129. The volumetric heat capacity (cal cm^{-3} $^{\circ}$C^{-1}) of soil mineral fraction is

    (a) 0.598   (b) 0.50   (c) 0.477   (d) 0.46

130. The specific heat (cal gm^{-1} $^{\circ}$C^{-1}) of soil organic fraction is

    (a) 0.25   (b) 0.18   (c) 0.20   (d) 0.46

131. The specific heat (cal gm^{-1} $^{\circ}$C^{-1}) of soil mineral fraction is

    (a) 0.25   (b) 0.18   (c) 0.20   (d) 0.46

132. The specific heat (cal gm^{-1} $^{\circ}$C^{-1}) of soil solids is

    (a) 0.25   (b) 0.18   (c) 0.20   (d) 0.46

134. Volumetric heat capacity is obtained by multiplying the specific heat with

    (a) Bulk density of soil   (b) Particle density of soil

    (c) 1/bulk density of soil   (d) 1/ Particle density of soil

135. The unit of volumetric heat capacity is

    (a) cal gm^{-1} $^{\circ}$C^{-1} (b) cal cm^{-3} $^{\circ}$C^{-1}

    (c) gm cal $^{\circ}$C^{-1} (d) cal $^{\circ}$C^{-1}

136. The unit of thermal conductivity is

    (a) cal $^{\circ}$C^{-1} (b) cal sec^{-1} $^{\circ}$C^{-1}

    (c) cm^2 sec^{-1} (d) cal cm^{-1} sec^{-1} $^{\circ}$C^{-1}

137. The unit of thermal diffusivity is

    (a) cal $^{\circ}$C^{-1} (b) cal sec^{-1} $^{\circ}$C^{-1}

    (c) cm^2 sec^{-1} (d) cal cm^{-1} sec^{-1} $^{\circ}$C^{-1}

138. The most sensitive type thermometers which can over a wide temperature range from-150 $^{\circ}$F to 550 $^{\circ}$F is

    (a) Mercury thermometer   (b) Bimetallic thermometer

    (c) Thermocouple thermometer   (d) Thermisters

139. When the mean annual soil temperature is from 8 $^{\circ}$C to less than 15 $^{\circ}$C and the difference between mean summer and mean winter soil temperature is more than 5 $^{\circ}$C the temperature regimes will be

    (a) Cryic   (b) Frigid   (c) Mesic   (d) Thermic

140. Which of the following soil colour conditions indicate anaerobic condition?

    (a) Black   (b) Red-Yellow   (c) Brown   (d) Blue –gray

141. Munsell colour notation is described by
     (a) Hue
     (b) Hue and Value
     (c) Hue and Chroma
     (d) Hue, Value and chroma

142. In, Munsell colour system, the degree of lightness of a colour in relation to a neutral gray scale is referred by
     (a) Hue
     (b) Value
     (c) Chroma
     (d) Neutral

143. In Munsell colour system, the relative purity of spectral colour is refered by
     (a) Hue
     (b) Value
     (c) Chroma
     (d) Neutral

144. The value of Munsell notation 10 YR3/6 is
     (a) YR
     (b) 10
     (c) 3
     (d) 6

145. The neutral colour (N) have
     (a) No hue
     (b) No chroma
     (c) No hue and zero value
     (d) No hue and zero chroma

146. Bright (High chroma) colors throughout the profile are the symptoms of
     (a) Well drained soils
     (b) Imperfectly drained soils
     (c) Poorly drained soils
     (d) Water logged condition during major part of the year

147. The colour of water logged soils is
     (a) Gray
     (b) Yellow
     (c) Brown
     (d) Black

148. With respect to the colour, the least productive soil has
     (a) Red colour
     (b) Gray colour
     (c) Yellow colour
     (d) White colour

149. With respect to the colour, the most productive soil has
     (a) Black colour
     (b) Brown colour
     (c) Red colour
     (d) Gray colour

150. The dominant spectral colour related to the wave length of light in Munsell soil colour chart is called
     (a) Hue
     (b) Value
     (c) Chroma
     (d) Neutral

151. In soil colour determination by Munsell soil colour chart hue is related to
     (a) Wavelength of light
     (b) Relative purity or strength of spectral colour
     (c) Degree of lightness or darkness of a colour
     (d) Intensity of light

152. In Munsell soil colour chart value number zero (0/) indicates of a colour
    (a) Pure white
    (b) Pure black
    (c) Pure yellow
    (d) Pure red

153. In Munsell soil colour chart value number ten (10/) indicates of a colour
    (a) Pure white
    (b) Pure black
    (c) Pure yellow
    (d) Pure red

154. In Munsell soil colour chart the value number ranges from
    (a) 0-8
    (b) 1-8
    (c) 0-10
    (d) 1-10

155. In Munsell soil colour chart the chroma number ranges from
    (a) 0-8
    (b) 1-8
    (c) 0-10
    (d) 1-10

156. In Munsell soil colour chart soil hue ranges from
    (a) 2.5 R to 10 Y
    (b) 5R to 10 Y
    (c) 0 R to 2.5 Y
    (d) 10 R to 5 Y

157. In Munsell soil colour chart hue ranges from
    (a) 0-8
    (b) 1-8
    (c) 0-10
    (d) 1-10

158. Kinetic energy expressed in
    (a) Erge
    (b) Joules
    (c) Newton
    (d) Both a & b

159. In water the hydrogen and oxygen atoms are attached at an angle of ______ ^{0}
    (a) 106
    (b) 125
    (c) 105
    (d) 95

160. Hydrogen and oxygen in the water molecule are linked by
    (a) Hydrogen bond
    (b) Oxygen bond
    (c) Both a & b
    (d) None

161. Diameter of water molecule
    (a) 5.6 A^0
    (b) 1.94 A^0
    (c) 0.6 A^0
    (d) 2.64 A^0

162. Density of which form of water is more
    (a) Liquid
    (b) Ice
    (c) Vapour
    (d) All are same

163. Amount of energy required by water to change from solid to liquid state
    (a) 540
    (b) 80
    (c) 620
    (d) 520 cal / gm

164. Water exhibit maximum density at ________ ^{0}C
    (a) 10
    (b) 9
    (c) 4
    (d) 0

165. When water freezes it expands by _____ %
    (a) 10
    (b) 9
    (c) 4
    (d) 25

166. SI unit of surface tension
    (a) Newton/m
    (b) Erge/cm^2
    (c) Dynes/cm
    (d) All

167. Surface tension of water is
    (a) 79.2　　　　(b) 72.7　　　　(c) 27.7　　　　(d) 57 dynes/cm

168. As the temperature increases surface tension
    (a) Decreases　　　　　　　　(b) Increases
    (c) No change　　　　　　　　(d) Slightly increase

169. The capillary rise of water is faster in __________ texture soil
    (a) Fine texture soil　　　　　　(b) Medium texture soil
    (c) Coarse texture soil　　　　　(d) None

169. The freezing and melting point of water _______ ^{0}C
    (a) 4　　　　　(b) 9　　　　　(c) 0　　　　　(d) 100

171. As solute concentration in soil solution increases the freezing point
    (a) Decreases　　(b) Increases　　(c) No change　(d) None

172. The thawing point of water is a _______ ^{0}C
    (a) 10　　　　(b) 4　　　　(c) 0　　　　(d) 9

173. Among the following which one is/ are true
    (a) Evaporation is a slow process
    (b) Evaporation takes place surface of liquid only
    (c) Evaporation takes at all temperature
    (d) All

174. Which of the following is / are with respect to boiling point of water
    (a) As pressure decrease boiling point also decreases
    (b) Boiling point increase with increases in salt content
    (c) Boiling point increase with increase in salt content
    (d) All

175. The process of direct change of water from solid to gaseous state
    (a) Sublimation　　　　　　　(b) Latent heat
    (c) Vaporization　　　　　　　(d) None

176. Heat capacity of water is ______ cal/ g
    (a) 1　　　　　(b) 0.2　　　　(c) 0.5　　　　(d) 0.45

177. Specific heat of air is _________ cal/g
    (a) 0.925　　　(b) 1　　　　(c) 0.25　　　　(d) 0.171

178. The term pF given by
    (a) Schofield　　(b) Braggs　　(c) Stokes　　(d) Sorenson

179. pF indicates
    (a) Moisture content   (b) Hydrogen ions
    (c) Both a & b   (d) None

180. The moisture tension at MWH close to
    (a) 1   (b) 0   (c) 10   (d) 100

181. After how many hours of rains the soil will attain field capacity
    (a) 24 hr   (b) 48-72 hr   (c) 30hr   (d) 94 hr

182. pF value at FC
    (a) 2.5   (b) 4.2   (c) 10   (d) 4.8

183. The moisture tension at maximum water holding capacity _________ bars
    (a) 0   (b) 10   (c) 2.5   (d) 4.8

184. Capillary water held at tension of ________ bars
    (a) 1/3 - 31   (b) 15-20   (c) 0.1 – 0.3   (d) 1-15

185. Evaporation is an
    (a) Endothermic   (b) Exothermic
    (c) Isothermic   (d) Hyperthermic

186. Among the following maximum water holding capacity found in __________
    (a) Loamy sand   (b) Sandy loam   (c) Loam   (d) Silty loam

187. The soil water tension at hygroscopic coefficient __________ bars
    (a) 3   (b) 15   (c) 20   (d) 31

188. Maximum water content at wilting coefficient found in
    (a) Sand   (b) Sandy loam   (c) Loam   (d) All

189. At same amount of water the availability of water maximum at
    (a) Sandy soils   (b) Silty soils   (c) Clay soils   (d) None

190. Evolution of heat due to absorption of water is known as
    (a) Heat of wetting   (b) Heat of energy
    (c) Heat of solution   (d) All

191. Among the following which one is exothermic process
    (a) Heat of wetting   (b) Evaporation
    (c) Heat of energy   (d) All

192. Heat of wetting expressed in
    (a) g/kg   (b) cal/gram   (c) Both a &b   (d) None

193. pF value of oven dry soils
    (a) 4.5   (b) 8.9   (c) 10   (d) 7

194. Physical classification of water given by
     (a) Schofield   (b) Briggs   (c) Sorenson   (d) Rode wald

195. Available water ranges between _________ bars
     (a) 0.1 – 0.3   (b) 1-31   (c) 1/3-15   (d) 15 -30

196. Lower infiltration of water found in
     (a) Sandy soils   (b) Saline soils   (c) Sodic soils   (d) All

197. Air dry soil have moisture tension of
     (a) 15 bar   (b) 31 bars   (c) 1000 bars   (d) 10000 bars

198. Maximum availability of water found in
     (a) Heavy clay soils   (b) organic soils
     (c) Loamy soils   (d) Gravel water

199. When ions are hydrated large amount of energy is released into the solution this is known as
     (a) Heat of wetting   (b) Heat of solution
     (c) Both a & b   (d) None

200. Neutron probe estimates moisture in soil by detecting molecules of
     (a) Water   (b) oxygen   (c) hydrogen   (d) carbon

210. Cluster of water molecules are formed by linking two neighboring water molecules by
     (a) Covalent bond   (b) Coordinate bond
     (c) Ionic bond   (d) Hydrogen bond

211. Water is considered as good solvent because
     (a) Dipolar in nature   (b) High dielectric constant
     (c) Hydrogen bonding with other material
     (d) All of these

212. The viscosity of water
     (a) 1 centi poise   (b) 1 poise
     (c) 1 dynes / cm^2 (d) 10 dynes / cm

213. The dielectric constant of pure water
     (a) 72   (b) 78.25   (c) 90.2   (d) 75.2

214. Presence of soluble salts reduce the water availability due to '
     (a) Decrease of field capacity
     (b) Increase wilting coefficient
     (c) Decreases in wilting coefficient
     (d) Relatively more decreases in of field capacity than wilting coefficient

215. The source of neutrons in a neutron moisture meter

    (a) Radium – beryllium   (b) Berlyium – cesium

    (c) Radium – americium   (d) Cesium – uranium

216. Neutron scattering technique for moisture estimation given by

    (a) Van bevel   (b) stone   (c) Schofield   (d) Bukhigam

217. The source of gama rays in gama ray attenuation technique is

    (a) Radium   (b) Beryllium   (c) Cesium   (d) Uranium

218. Field capacity indicates

    (a) Upper level of super available water

    (b) Upper level of plant available water

    (c) Lower level of plant available water

    (d) Upper level of capillary water

219. Wilting coefficient signifies the

    (a) Upper level of super available water

    (b) Upper level of plant available water

    (c) Lower level of plant available water

    (d) Upper level of capillary water

220. Water molecules are adsorbed on soil particles by process of

    (a) Adhesion   (b) Cohesion   (c) Gravity force   (d) All

221. Maximum water holding capacity of soil is measured by

    (a) Soil moisture box   (b) Keens raczkowski box

    (c) Core sampler   (d) Psychrometer

222. Capillary water in soil remains in

    (a) Macropores   (b) Micro pores   (c) Both a& b   (d) None

223. Capillary raise of water due to the force of

    (a) Adhesion   (b) Cohesion

    (c) Both a & b   (d) Atmospheric pressure

224. With increase in solute content in soil water the vapour pressure

    (a) Decreased

    (b) Increased

    (c) Unchanged

    (d) Slightly increased and become constant

225. Formula of wilting coefficient

    (a) Hygroscopic coefficient/ 6.68

    (b) Moisture coefficient/1.8

    (c) Hygroscopic coefficient/ moisture coefficient

    (d) Both a & b

226. Major potential involved in movement of unsaturated soil

    (a) Matric potential   (b) Gravity potential

    (c) Vapour pressure   (d) All

227. Among the following conditions movement of water in all direction found in

    (a) Saturated flow   (b) Unsaturated flow

    (c) Vapour movement   (d) All

228. Direction of flow of water in unsaturated condition

    (a) Downward   (b) Lateral

    (c) All direction   (d) both a & c

229. Consider the following which one is /are true

    (a) In unsaturated flow all pores filled with water

    (b) In saturated flow only micro pores are filled with water

    (c) In vapour movement all pores are filled with air

    (d) All

230. Concept of soil water potential is given by

    (a) Briggs   (b) Shantz   (c) Bukingham (d) Schofied

231. 1 bar is equal to

    (a) 106 dynes   (b) 0.9869 atm   (c) 1020 cm of water column

    (d) All

232. In the saturated soils the matric potential is

    (a) 0   (b) 1.2   (c) 2.5   (d) 4.2

233. pF indicates

    (a) Moisture content   (b) Hydrogen ions

    (c) Osmotic potential   (d) None

234. Tensiometer measures

    (a) Total soil water potential   (b) Matric potential

    (c) Osmotic potential   (d) Both A & C

235. Entry of water into soil is called
   (a) Percolation
   (b) Infiltration
   (c) Permeability
   (d) Hydraulic conductivity

236. Matric potential is mainly due to
   (a) Salts
   (b) Gravity
   (c) Adsorptive and capillary force
   (d) All

237. Moving force for water flow in a saturated soils is the gradient of
   (a) Positive
   (b) Negative
   (c) Neutral
   (d) Pressure potential

238. Moving force for water flow in a unsaturated soils is the gradient of
   (a) Positive   (b) Negative   (c) Neutral   (d) Pressure

239. In unsaturated condition maximum flow of water found in
   (a) Sandy loam   (b) Clay loam   (c) Silty loam   (d) Silty clay

240. In saturated condition maximum flow of water found in
   (a) Sandy loam   (b) Clay loam   (c) Silty loam   (d) Silty clay

241. Soil moisture characteristic curve shows the relationship between
   (a) Matric potential and volumetric water content
   (b) Osmotic potential and volumetric water content
   (c) Matric potential and osmotic potential
   (d) All

242. In the phenomena of hysteresis more energy required for
   (a) Sorption
   (b) Desorption
   (c) Equal to both the process
   (d) All

243. The curves obtained due to difference in process of sorption and desorption
   (a) Hysteresis
   (b) Sorption
   (c) Absorption curve
   (d) All

244. Water held by the soil solids at the suction of 31 bars is termed as
   (a) Available water
   (b) Seepage water
   (c) Drainage water
   (d) Hygroscopic water

245. In moist range unsaturated conductivity of soils is in the sequence of
   (a) Sand < Loam < Clay
   (b) Clay <Loam < Sand
   (c) Sand>Loam > Clay
   (d) None of these

246. The volume of rewetted soils is greater than original soil volume due to
   (a) Swelling   (b) Shrinking   (c) Run off   (d) Drying

247. Tensiometer works upto a tension of
   (a) 1.0 bar   (b) 1.5 bar   (c) 0.85 bar   (d) 1.25 bar

248. The formula for calculating height of water rise in a soil is
   (a) h = 0.3/d   (b) h=0.15/d   (c) 0.25/d   (d) 0.35/d

249. Generally the water content at wilting point is more in
   (a) Loamy soil   (b) Sandy soils
   (c) Clayey soil   (d) Sandy loam soil

250. As Viscosity of water increases , hydraulic conductivity of soil
   (a) Increases   (b) Decreases
   (c) Remains static   (d) None of these

251. FOURIER' Law states
   (a) Water flow through soil   (b) Vapour flow through soil
   (c) Heat flow through soil   (d) None of these

252. Logarithm of the negative pressure head in centimeter of water column height is known as
   (a) $_pH$ (b) E^h (c) P^F (d) ET

253. The moving force of water in a saturated soils is the gradient of
   (a) Gravitational potential   (b) Positive pressure potential
   (c) Osmotic potential   (d) Matric potential

254. Pressure plate apparatus used form measurement of soil moisture upto
   (a) 1 bar   (b) 0.85 bar   (c) 15 bar   (d) None

255. Which among the following is related to movement of water
   (a) Darcy's law   (b) Ficks law
   (c) Ohms law   (d) Fourier law

256. Consider the following which is / are true
   (a) Soil moisture content decreases with increase in soil moisture tension
   (b) Soil moisture content increases with increase in soil moisture tension
   (c) Soil moisture content doesn't
   (d) All

257. As the temperature increases the hydraulic conductivity
   (a) Increases   (b) Decreases
   (c) No change   (d) None

258. Tensiometer measures
   (a) Matric potential   (b) Osmotic potential
   (c) Pressure potential   (d) All

259. Solute potential known as
    (a) Matric potential   (b) Osmotic potential
    (c) Pressure potential   (d) All

260. Water potential is the sum of
    (a) Matric potential and osmotic potential
    (b) Osmotic potential and pressure potential
    (c) Matric potential, osmotic potential and gravitational potential
    (d) None

261. Very poor infiltration found in
    (a) Acid soils   (b) Saline soils
    (c) Saline alkali   (d) Alkali soils

TRUE OR FALSE

1. Arrangement of secondary soil particles and their aggregation into some definite shape termed as soil structure

2. Aggregation of sand silt clay is called soil structure

3. Clod is naturally formed soil mass

4. Soil structure is permanent property of soil and cannot be altered

5. All plant growth factors are influenced by soil structure

6. Soil structure has a pronounced effect on erodibility of soil

7. Dispersion of clay is necessary for formation of good soil aggregation

8. In sandy soils roots and fungi hyphae are mainly responsible for aggregate formation

9. Soil under grassland has high state of aggregation

10. Soils which are high in aeration status have high infiltration, percolation and water retention capacity

11. Platy type of soil structure is best for plant growth

12. Spheroidal type structure commonly found in surface soils

13. Wetting and drying affect soil aggregation

14. The cations on soil colloids acts as a cementing agent

15. Soil with very low clay content have tendency to form strong aggregates

16. Laterite soils have stable aggregates

17. Puddling improves the soil structure

18. Branched root system improves the soil structure

19. Phosphate anion play important role in soil aggregation

20. Granular structure is a simple structure

21. The particle size of the mineral soil changes readily

22. Texture of soil cannot alter easily

23. Soil separates are also known as fine earth

24. Coarse fragments are also considered for determination of textural classes

25. Coarse fragments are bad conductor of heat

26. Coarse fragments in soil make the soil less permeable

27. Soils dominated by sand fractions have low water holding capacity and posses good drainage and aeration

28. Sandy soils have high specific surface area as the particles size more

29. Sandy soils have no CEC

30. Clay fractions of soil are highly sticky and low plastic in nature

31. Clay particles are organic in nature

32. Clay particles have high swelling and shrinkage property

33. Specific surface area of clay particles is very high

34. Clay particles are composed of aluminosilicates

35. Silt is the most chemically active part of soil

36. Sand silt and clay particles can exhibit the flocculation and defloculation phenomena

37. Soils which have low chemical activity they are resistance to weathering

38. Hydrometer works on the principle of density

39. Stokes law is valid for platy shaped particles

40. Clay is a primary soil particles

41. Light texture soils are more susceptible to compaction

42. Ethylene glycerol method used to measure specific surface area of soil particles

43. Water molecules is electrically neutral

44. The covalent bond linkage in water molecule is significantly weaker than hydrogen bond

45. Ice floats on water mainly because of less density of ice compare to water

46. Changes of water from one form to another result in changes of energy

47. The degree of hydrogen bonding constant in water molecules

48. Surface tension of water is very low as compare to other liquids

49. The height of capillary raise of water decreases with increase in temperature

50. Water molecules possess both positive and negative poles

51. The viscosity of water increase with increase in temperature

52. Evaporation is an endothermic reaction

53. Boiling point increases with increase in pressure

54. Condensation is endothermic reaction

55. Soil water content governs the air content & gas exchange of the soil

56. Ice has a less kinetic energy than liquid water

57. The dissolved solutes increases the free energy of water

58. Water pressure decreases by presence of solutes

59. Near the particle surface the viscosity of water is very high

60. Capillary raise of water is faster in sandy soils

61. Water has low heat capacity

62. The attraction for solids for water decreases the free energy of soil water

63. Sandy soils hold more water than clay soils

64. At field capacity most of water is retained in macro pores

65. Gravitational water movement is due to capillary force

66. The value of maximum water holding capacity is equal to the total porosity

67. In saturated condition sandy soils attain field capacity quicker than clay soils

68. All capillary water is available to plants

69. The available water content increases with increase in fineness texture

70. Gravitational water is not available to plant

71. Increase in organic matter content in sandy soils decreases the water holding capacity

72. Field capacity value of loamy soils is more than the sandy soils

73. As the moisture content increases the surface tension decreases

74. In saturated soils all pores are filled with water

75. In unsaturated soil all pores are filled with air

MATCH THE FOLLOWING

1.

	A		B
1	Black plastic mulch	a	Thermal insulators
2	Paper and straw mulch	b	Reduce outgoing temperature
3	Aluminum foil mulch	c	Sharply increase outgoing temperature
4	Transparent plastic mulch	d	Increase outgoing temperature
5	Opaque mulch	e	Green house effect

2.

	A		B
1	Soil crust	a	Plastic limit
2	Application of SSP or gypsum	b	Plastic number
3	Montmorillonite clay	c	Liquid limit
4	Kaolinite clay	d	Natural aggregation
5	Type	e	Thin crust
6	Class	f	Thick & hard clay
7	Peds	g	Control of soil crust
8	Lower plastic limit	h	Penetrometer
9	High plastic limit	i	Shape & arrangements of peds
10	Hydration energy	j	Size of peds

3.

	A (Particle)		B (ISSS) (mm)		USDA (mm)
a	fine sand	e	0.02 – 0.002	1	1.0 – 0.5
b	silt	f	0.2 – 0.02	2	< 0.002
c	coarse sand	g	<0.002	3	0.05 – 0.002
d	clay	h	2.0 – 0.2	4	0.25 – 0.1

UNITS

1. Surface area : **m²/g**
2. Bulk Density : **Mg m⁻³ or g cm⁻³ or g/cc**
3. Specific volume : **cm³/g**
4. Viscosity : **g sec cm⁻¹ or Poise**
5. Gravity : **m/sec² (or) N/kg**
6. Hydraulic conductivity : **cm sec⁻¹**
7. Soil water potential or mass basis : **Joules kg⁻¹**
8. Soil water potential on weight basis : **cm or m**
9. Thermal diffusivity : **cm²/sec**
10. Specific heat : **Cal/g**
11. Thermal conductivity : **J/°C/cm/sec**
12. Volumetric heat capacity : **J/m³/°C**
13. CEC : **Cmol(p⁺)kg⁻¹**
14. C-axis : **A°**
15. RSC : **me L⁻¹**
16. ODR : **g/cm²/min**
17. EC : **µ mho/cm or dSm⁻¹**

FORMULAS

1. Relation between porosity (f) and avoid ratio (e) _____________ **[e= f/(1-f)]**
2. Relation between porosity (f), bulk density (pb) and particle density (ps) ____________ **[f=1-pb/ps]**
3. Relation between mass wetness (w) and volume wetness (θ) _____________ **(w = θ.pw/pb)**
4. Relation between volume wetness (θ), fractional air content (fa) and degree of saturation(s) _____________ **(θ = F - Fa)**
5. Relation between volume wetness (θ) and degree of saturation(s) and porosity (f)________ **(F = θ/s)**
6. Relation between air filled porosity (fa), porosity f) and degree of saturation (s)_______ **[fa= f(1-s)]**
7. Stokes law__________________ $\eta \left[V = \dfrac{2\left(p_s - p_w\right)gr^2}{9\eta} \right]$
8. Percentage of total pore space _________________ **[1-pb/ps] × 100**

KEY TO FILL IN THE BLANKS

1. Soil texture
2. Basic/permanent
3. Soil texture
4. Soil skeleton
5. Robinson pipette method
6. Pipette
7. 2 mm
8. Coarse or rock fragments
9. Texture
10. Texture
11. 7
12. 12
13. 12
14. 0.05 mm
15. 0.05 to 0.002 mm
16. < 0.002 mm
17. < 0.002 mm
18. Gravel
19. Feel
20. Mechanical analysis
21. H_2O_2
22. Sodium hexa Meta Phosphate
23. Sedimentation
24. 10-20 % clay
25. Sandy
26. Loamy soils
27. Stokes law
28. 2.65 g cm^{-3} or Mg m^{-3}
29. Viscosity
30. Hydrometer method
31. International pipette method
32. (Radius)^2 of the particle and Viscosity
33. Clay
34. Soil structure
35. Size
36. Silt
37. Poor
38. 10,000
39. Sandy
40. Clay
41. Zeta potential
42. Clay
43. aggregation
44. Illite
45. Crumby type
46. Granular
47. Platy
48. Surface soils
49. Block type
50. Laterite
51. Hydrous oxides of iron and Al
52. Wet sieving method
53. Platy structure
54. Gravimetric water content
55. Quartz
56. Sand and silt
57. Bouyant force
58. Sand, silt and clay
59. $\frac{4}{3}\pi r^3$
60. 20 microns
61. Clay
62. 0-15 cm
63. 2.24 × 10^6 kg ha^{-1}
64. Soil structure
65. Type, Class and Grades
66. 4

67. Platy
68. Laminar
69. Platy
70. prism
71. Prismatic structure
72. Columnar structure
73. Blocky structure
74. Angular Blocky structure
75. Crumb structure
76. Granular structure
77. Type
78. Block
79. Soil consistency
80. Plasticity
81. Friable
82. Plasticity index
83. Cohesion
84. Adhesion
85. Shrinkage limit
86. Splashing affect of rain drop
87. Red sandy loam
88. Silty clay loam
89. Clay soils
90. Broken peds
91. Peds
92. Ped
93. Weak soils
94. Moderate soils
95. Strong soils
96. Aggregates
97. Clod
98. Fragment
99. Concretion
100. Type
101. Class
102. Compound structures
103. Crumb
104. Granular
105. Angular blocky
106. Sub-angular blocky
107. < 1c.m
108. Crumb
109. Shape and arrangements
110. Size
111. Grade
112. Grade
113. Simple structure
114. Sandy soils
115. Paddy soils
116. Vesicular or honeycomb structure
117. Laterite soils
118. B horizons
119. Create platy structures
120. Durability
121. Sodium
122. Platy
123. Rice
124. Angular Blocky
125. Flocculation
126. Flocculation and Cementation
127. Ca
128. Ca
129. Na
130. Ca
131. Deflocculation
132. Organic matter
133. soil conditioners
134. Gypsum
135. N and P fertilizers
136. Clay
137. Legumes
138. Crumb

139. E.J. Russel	140. Phosphate
141. Salt affected soils	142. Particle density
143. Soil Air	144. Remains unaltered
145. 2.65 Mg/m^3	146. Bulk density
147. Apparent Specific Gravity	148. True density
149. Decreases	150. Compaction
151. Penetrometer	152. Sodium
153. Soil crusting	154. Clay
155. Hard crust	156. Kaolinite
157. Soil crust	158. Strength of soil crust method test
159. Soil consistency	160. Plasticity
161. Upper plastic limit	162. 15 %
163. Higher is the plastic number	164. Plastic number
165. Plasticity	166. Moist
167. Clay	168. Plasticity
169. Increase the plastic number	170. Constant
171. Coefficient of linear expansion	172. Black soil
173. Vertisols	174. $\{[L_M\text{-}L_d]/L_d\} \times 100$
175. Pore space	176. Porosity
177. Air filled porosity	178. Air filled porosity
179. Void Ratio	180. Void Ratio
181. Volume wetness	182. Degree of saturation
183. Bulk density	184. Clayey soil
185. < 0.08 mm	186. Increases
187. Bulk density	188. Bulk density
189. 55	190. 1.35 g/cm^3
191. 40 %	192. 1.1 -1.3 g Cm^3
193. 1.4 -1.8 g Cm^3	194. 0.5 g Cm^3
195. Specific volume	196. Core method and excavation method
197. Oxides of iron	198. Chroma
199. Hue	200. Value
201. 3A°	202. 105 °C
203. 80 Cal	204. 1.013×10^6 dynes cm^2 or 1.013 bar or 101.3 kPa
205. 10	206. 1/10
207. 100	208. 1020 cm
209. 106 dynes /cm^2	210. 1000 cm of water column

211. 4 ℃	212. Zero
213. Zero	214. -1000 bar
215. 1/3 bar or 33 kPa	216. 15 bar or 1500 kpa
217. 10-33 kPa	218. 33-1500 kpa
219. -31 bars	220. -1000 bars
221. -0.85 bars	222. 0.10 bars
223. 1.8×10^{-14}	224. 1.01×10^{-14}
225. Negative	226. Positive
227. 105 ℃	228. Infiltration
229. Porosity	230. Sandy soils
231. Unity or One	232. $Q = ki$
233. Fourth power	234. Drainage
235. 0.3/ d	236. Decreases
237. Cohesive and Adhesive	238. Cohesion and Adhesion
239. Osmotic potential	240. Sandy
241. Sands> Loams > Clay	242. Clay > Loams > Sands
243. Schofield	244. Decreases
245. Radius	246. Hydraulic conductivity
247. cm/min	248. Ca soil > K soil > Na soil
249. Matric potential	250. Matric potential
251. Gravitational potential	252. Solute and pressure potential
253. Available water	254. Drainable water
255. Spherical	256. Hydrogen bonding
257. Solid , liquid , gas	258. Directly
259. Saturated , dry soils	260. Cooling
261. Cohesion	262. Matric potential
263. Field capacity	264. Four
265. 3.4×10^{22}	266. 4^0 c
267. 4^0 C	268. 10^{14}
269. Cohesion and adhesion	270. Organic matter
271. 1/3-15	272. Field capacity
273. Capillary water	274. $105\ ^0$C
275. Meniscus	276. Concave
277. Surface tension	278. Field capacity
279. Lower	280. 1-15
281. Capillary water	282. Micropores

283. Hygroscopic water
284. Oven drying
285. Adhesion
286. Gravity
287. Aeration porosity
288. Clay > Loam > Sandy
289. Black > Alluvial > Laterite
290. Heat of solution
291. Heat of wetting
292. Positive
293. Negative
294. m+ p+ s +g
295. 580 cal
296. one
297. 2.53
298. 7.0
299. 4.18
300. 4.50
301. Tensiometer
302. Tensiometer
303. 800-900 cm
304. 0.85 bars
305. Tensiometer
306. Porus cup apparatus and pressure plate apparatus
307. 1 bar
308. 15 bars
309. Neutron probe method
310. Electrical resistance method, Neutron scattering method and Gamma ray attenuation technique
311. Gypsum
312. Cesium137
313. BF_3
314. Moisture stability
315. Zero
316. Hydraulic gradient
317. Sand < Silt< Clay
318. Matric potential or –ve pressure potential
319. Air
320. Gravitational force
321. Adsorption
322. Thermal conductivity
323. Sandy soils
324. Albedo
325. Heat capacity /thermal capacity
326. Thermal conductivity
327. Sand>loam>clay>peat
328. Thermal diffusivity
329. Thermal retentivity
330. Thermal capacity
331. Conduction
332. Green house effect
333. Retentivity
334. White succulent roots with less branching
335. 0.2 cal/g
336. Organic matter
337. Sandy > Loam > Clay > Peat
338. Conduction
339. Sun
340. 10 °C
341. Thermal diffusivity
342. 0.50%
343. CO_2
344. Diffusion
345. Diffusion

346. Radiation   347. 20 %
348. 20 -40 $\times 10^{-8}$ g/cm^2/min   349. Oxygen content
350. oxygen   351. Hillel .D

KEY TO CHOOSE THE CORRECT ANSWER

1. c	2. a	3. c	4. c	5. d	6. c
7. b.	8. a	9. b	10. d	11. a	12. a
13. a	14. a	15. a	16. d	17. a	18. c
19. a	20. c	21. b	22. b	23. c	24. a
25. a	26. c	27. b	28. c	29. a	30. b
31. b	32. a	33. d	34. b	35. a	36. d
37. a	38. a	39. d	40. a	41. b	42. a
43. c	44. d	45. a	46. b	47. a	48. d
49. d	50. c	51. b	52. b	53. c	54. a
55. b	56. b	57. d	58. a	59. a	60. c
61. a	62. d	63. b	64. a	65. b	66. d
67. d	68. d	69. c	70. a	71. b	72. a
73. b	74. a	75. d	76. d	77. c	78. b
79. d	80. a	81. c	82. d	83. b	84. b
85. d	86. c	87. d	88. d	89. a	90. d
91. c	92. a	93. a	94. c	95. b	96. a
97. c	98. b	99. b	100. b	101. d	102. b
103. c	104. c	105. a	106. c	107. a	108. a
109. c	110. a	111. c	112. b	113. a	114. b
115. d	116. b	117. a	118. d	119. d	120. b
121. d	122. c	123. c	124. c	125. a	126. c
127. c	128. c	129. c	130. d	131. b	132. c
134. b	135. b	136. d	137. c	138. d	139. c
140. d	141. d	142. b	143. c	144. c	145. d
146. a	147. a	148. d	149. a	150. a	151. a
152. b	153. a	154. c	155. a	156. d	157. c
158. d	159. c	160. a	161. d	162. a	163. b

164. c	165. b	166. a	167. b	168. a	169. c
170. a	171. a	172. c	173. d	174. d	175. a
176. a	177. d	178. a	179. a	180. b	181. b
182. a	183. a	184. a	185. a	186. d	187. d
188. c	189. a	190. a	192. b	193. d	194. b
195. c	196. c	197. c	198. b	199. b	200. c
210. d	211. d	212. a	213. b	214. b	215. a
216. b	217. c	218. b	219. c	220. a	221. b
222. b	223. c	224. a	225. d	226. a	227. c
228. b	229. d	230. c	231. d	231. a	232. a
233. a	234. b	235. b	236. c	237. a	238. b
239. d	240. a	241. a	242. b	243. a	244. d
245. a	246. a	247. c	248. a	249. c	250. b
251. c	252. c	253. b	254. c	255. a	256. a
257. a	258. a	259. b	260. c	261. d	

KEY TO TRUE OR FALSE

1. False [True: Primary soil particles]
2. True
3. False [True: Artificially formed]
4. False [True: Not a permanent property and can be changed]
5. True
6. True
7. False [True: Flocculation]
8. True
9. True
10. False [True: Low water retention]
11. False [True: Crumby structure]
12. True
13. True
14. False [True: Acts as a flocculating agent]
15. False [True: It form weak aggregation]
16. True
17. False [True: Destroy the soil structure]
18. True
19. True
20. False [True: Compound structure]
21. False [True: Doesn't change readily]
22. True
23. True

24. False [True: Not considered because they are not a part of soil]

25. False [True: Good conductor]

26. False [True: Make more permeable]

27. True

28. False [True: SSA is inversely proportion to size]

29. True

30. False [True: Highly sticky and highly plastic]

31. False [True: Inorganic]

32. True

33. True

34. True

35. False [True: Clay]

36. False [True: Only clay can exhibit]

37. True

38. True

39. False [True: Valid for spherical shape]

40. True

41. False [True: Less susceptible]

42. True

43. True

44. False [True: Stronger]

45. True

46. True

47. False [True: It varies with temperature]

48. False [True: Very high]

49. True

50. True

51. False [True: Decreases]

52. True

53. True

54. False [True: Exothermic because heat is evolved for condensation of water vapour]

55. True

56. True

57. False [True: Solutes lower the free energy]

58. True

59. True

60. True

61. False [True: High heat capacity]

62. True

63. False [True: Hold less water]

64. False [True: Retained in micro pores]

65. False [True: Gravitational force]

66. True

67. True

68. False [True: Only part of capillary water available to plant 1/3 -15 bar]

69. False [True: Increases only up to silty loam texture beyond which it decreases with increase in fineness]

70. True

71. False [True: It increases the water holding capacity]

72. True

73. True

74. True

75. False [True: Only part of soil filled with air remaining filled with water]

KEY TO MATCH THE FOLLOWING

1.	2-a	4-b	3-c	5-d	1-e
2.	8-a	10-b	9-c	7-d	4-e
	3-f	2-g	1-h	5-i	6-j
3.	af4	be3	ch1	dg2	

<table><tr><td>3</td><td>

SOIL FERTILITY AND NUTRIENT MANAGEMENT
</td></tr></table>

Hamsa N. and Prabhavathi N.

CHOOSE THE CORRECT ANSWER

1. Organic matter increases at the temperature
   (a) 0 to 35 °C
   (b) 20- 25 °C
   (c) > 35 °C
   (d) 60 °C

2. White bud of maize is due to the deficiency of
   (a) Copper
   (b) Zinc
   (c) Boron
   (d) Sulphur

3. Molybdenum is a component of
   (a) Chlorophyll
   (b) Nitrate reductase
   (c) Urease
   (d) Nitrogenase

4. Nutrient index concept is used in the preparation of
   (a) Soil fertility map
   (b) Soil productivity map
   (c) Both a and b
   (d) Nutrient diagnosis chart

5. Father of soil testing
   (a) Dockucheiv
   (b) Troug
   (c) Jackson
   (d) None of the above

6. The anion responsible for soil aggregation is
   (a) Phoshate
   (b) Nitrate
   (c) Sulphate
   (d) Carbonate

7. The term 'functional nutrient' was introduced by
   (a) Arnon and Stout
   (b) Epstein
   (c) Jackson
   (d) Nicholas

8. Activity of acetyl coenzyme is promoted by
   (a) Phosphorus
   (b) Potassium
   (c) Zinc
   (d) Molybdenum

9. Ultra micro nutrient element is
   (a) Mn
   (b) Mg
   (c) Mo
   (d) Si

10. Order of solubility of humic substances
    (a) Humic acid > Fulvic acid> Humin
    (b) Fulvic acid<Humic acid > Hum
    (c) Fulvic acid>Humic acid > Humin
    (d) Fulvic acid<Humic acid <Humin

11. Ascending order of molecular mass of humic substances
    (a) Humic acid > Fulvic acid> Humin
    (b) Fulvic acid<Humic acid > Humin
    (c) Fulvic acid>Humic acid > Humin
    (d) Fulvic acid<Humic acid <Humin

12. Mitscherlich equation is
    (a) $dy/dx = c(A-y)$ (b) $dy/dx = c(A+y)$
    (c) $dy/dx = (A-y)/c$ (d) All the above

13. Tracer based techniques in soil fertility evalution methods are
    (a) A- value   (b) E- value   (c) L- value   (d) All the above

14. Sulphur bearing vitamin
    (a) Biotin   (b) Riboflavin   (c) Thiamine   (d) Both a and c

15. Zn is involved in the synthesis of
    (a) Biotin   (b) Riboflavin   (c) Auxin   (d) Gibberlin

16. Freckling in sugarcane is due to deficiency of
    (a) Zn   (b) B   (c) Mn   (d) Si

17. Boll shedding in cotton is due to the deficiency of
    (a) B   (b) Ca   (c) Zn   (d) Mn

18. The element that is considered as detoxifier of Ni is
    (a) Cu   (b) Ca   (c) Co   (d) Mg

19. Antagonistic interaction exists among the nutrients
    (a) P and Zn   (b) Ca and P   (c) Only a   (d) Both a and b

20. Range of nutrient content in plants associated with optimum crop yields is called
    (a) Deficient   (b) Insufficient   (c) Sufficient   (d) Critical

21. Extent of inorganic form of nitrogen in soil is
    (a) 48 %   (b) 2 %   (c) 98 %   (d) 70 %

22. Extent of organic form of nitrogen in soil is
    (a) 48 %   (b) 2 %   (c) 98 %   (d) 70 %

23. The amount of N present in Indian soil
    (a) 0.02-0.44 %   (b) 1-2 %   (c) 95-98 %   (d) 50-70 %

24. Among the following, which one is a purely a chemical process?
    (a) Nitrogen fixation   (b) Denitrification
    (c) Ammonia volatilization   (d) None of these

25. The examples for non symbiotic N fixation
    (a) Blue green algae   (b) Nostoc
    (c) Anabeana   (d) All of these

26. Major steps involved in the mineralization process
    (a) Aminization   (b) Ammonification
    (c) Nitrification   (d) All of these

27. What is the optimum temperature required for nitrification
    (a) 20 to 30!   (b) 45 to 50!
    (c) 25 to 40!   (d) All of the above

28. The accessory structural elements of living tissue
    (a) N,P & K   (b) N,P & S
    (c) Ca, Mg & S   (d) Ca,Mg & K

29. Effect of increase in application of nitrogen fertilizers in plant
    (a) Lodging   (b) susceptible to pest and disease
    (c) Early maturity   (d) Both a and b

30. P fixation occurs in soil with
    (a) Only acidic pH   (b) Only alkaline pH
    (c) Both acidic and alkaline pH   (d) Neutral pH

31. The total content of P in Indian soils
    (a) 100 to 2000ppm   (b) 0.1- 0.5 %
    (c) 10-20 %   (d) 10000-20000 ppm

32. Organic form of P
    (a) Inositol P   (b) Phospholopids
    (c) Nucleic acids   (d) All of these

34. Best method of application of P fertilizers to soil
    (a) Broadcasting   (b) Band placement
    (c) Fertigation   (d) Row application

35. P fixation demonstrated by
    (a) Sprengel   (b) Lewis
    (c) Thomas way   (d) Bray and Olsen

36. P solubilizing microorganisms
    (a) Aspergillus   (b) Bacillus   (c) Only a   (d) Both a and b

37. Which element is important for root growth
    (a) P   (b) N   (c) K   (d) Fe

38. Which element is energy currency in plant
    (a) P (b) N (c) K (d) Fe

39. The element which is not a part of any cell organelle
    (a) P (b) N (c) K (d) Fe

40. The amount of readily available K in soil
    (a) 200 ppm (b) 20 % (c) 0.2 % (d) 2 %

41. Q/I relationship was first given to the nutrient
    (a) P (b) N (c) K (d) Fe

42. The element Example for luxury consumption
    (a) P (b) N (c) K (d) Fe

43. The availability of the Zn in soil will be affected by the over application of
    (a) P (b) S (c) K (d) B

44. The element which is essential for cell wall membrane structure and increase stiffness of plant
    (a) N (b) Ca (c) K (d) Mg

45. The element essential for translocation of carbohydrates
    (a) N (b) Ca (c) K (d) Zn

46. Dieback symptom is typical deficiency symptom of
    (a) N (b) Ca (c) K (d) Zn

47. The main role of molybdenum
    (a) N fixation (b) Perroxidase activity
    (c) Nitrate reductase activity (d) Co enzyme A synthesis

48. The main role of boron
    (a) Pollen formation (b) Amino acid synthesis
    (c) Nitrate reductase activity (d) Co enzyme A synthesis

49. Fixation of nitrogen in root nodules is facilitated by
    (a) N (b) Mo (c) Mn (d) Mg

50. The nutrient that is a part of plastocyanin protein is
    (a) Ni (b) Co (c) Zn (d) Cu

51. The element that is subjected to highly leaching losses
    (a) N (b) S (c) Zn (d) Ca

52. Which is the nitrification inhibitor
    (a) Nitrapyrin (b) Oxamide
    (c) Dicyandiamide (d) a and c

53. Example of slow release N fertilizer
    (a) Nitrapyrin   (b) Oxamide
    (c) Dicyandiamide   (d) a and c (a)

54. The nutrient element involved in transformation of sugar to starch is
    (a) P   (b) K   (c) Zn   (d) Ca

55. The nutrient element involved in transfer of genetic characteristics from one generation to the next is
    (a) Ni   (b) Co   (c) Zn   (d) P

56. The plants shows deficiency symptoms when the concentration of P in tissues is less than
    (a) 50000 ppm   (b) 1000 ppm   (c) 10000 ppm (d) 5000 ppm

57. Bronze colouration of older leaves is the deficiency symptom of
    (a) N   (b) Ca   (c) K   (d) P

58. The nutrient element necessary for water transport in plants is
    (a) K   (b) Ca   (c) Zn   (d) P

59. The plants shows deficiency symptoms when the concentration of calcium in tissues is less than
    (a) 50000 ppm   (b) 1000 ppm   (c) 10000 ppm (d) 5000 ppm

60. The plants shows deficiency symptoms when the concentration of magnesium in tissues is less than
    (a) 50000 ppm   (b) 1000 ppm   (c) 10000 ppm (d) 5000 ppm

60. The plants shows deficiency symptoms when the concentration of manganese in tissues is less than
    (a) 50 ppm   (b) 100 ppm   (c) 25 ppm   (d) 250 ppm

61. The plants shows deficiency symptoms when the concentration of zinc in tissues is less than
    (a) 15 ppm   (b) 100 ppm   (c) 25 ppm   (d) 50 ppm

62. The plants shows deficiency symptoms when the concentration of zinc in tissues is less than
    (a) 15 ppm   (b) 50 ppm   (c) 25 ppm   (d) 5 ppm

63. The plants shows deficiency symptoms when the concentration of molybdenum in tissues is less than
    (a) 1.5 ppm   (b) 0.1 ppm   (c) 0.5 ppm   (d) 1.0 ppm

64. The plants shows deficiency symptoms when the concentration of nickel in tissues is less than
    (a) 1.5 ppm   (b) 0.1 ppm   (c) 0.5 ppm   (d) 1.0 ppm

65. The plants shows deficiency symptoms when the concentration of chlorine in tissues is less than

(a) 15 ppm   (b) 1.0 ppm   (c) 100 ppm   (d) 10 ppm

66. The plants shows deficiency symptoms when the concentration of selenium in tissues is less than

(a) 1.5-2.5 ppm   (b) 1-5 ppm   (c) 10-20 ppm   (d) 0.1-0.3 ppm

67. The average concentration of hydrogen in plant tissues is

(a) 1.5 %   (b) 1 %   (c) 10 %   (d) 6 %

68. The average concentration of oxygen in plant tissues is

(a) 45 %   (b) 15 %   (c) 10 %   (d) 60 %

69. The average concentration of carbon in plant tissues is

(a) 45 %   (b) 15 %   (c) 10 %   (d) 60 %

70. The average concentration of nitrogen in plant tissues is

(a) 4.5 %   (b) 1.5 %   (c) 1.0 %   (d) 6.0 %

71. The average concentration of phosphorus in plant tissues is

(a) 0.4 %   (b) 2.0 %   (c) 1.0 %   (d) 1.5 %

72. The average concentration of magnesium in plant tissues is

(a) 0.4 %   (b) 2.0 %   (c) 1.0 %   (d) 1.5 %

73. The average concentration of potassium in plant tissues is

(a) 0.4 %   (b) 2.0 %   (c) 1.0 %   (d) 1.5 %

74. The average concentration of calcium in plant tissues is

(a) 0.5 %   (b) 2.0 %   (c) 1.0 %   (d) 1.5 %

75. The average concentration of sulphur in plant tissues is

(a) 0.5 %   (b) 0.1 %   (c) 1.0 %   (d) 1.5 %

76. The average concentration of chlorine in plant tissues is

(a) 15 ppm   (b) 1.0 ppm   (c) 100 ppm   (d) 10 ppm

77. The average concentration of iron in plant tissues is

(a) 15 ppm   (b) 1.0 ppm   (c) 100 ppm   (d) 10 ppm

78. The average concentration of boron in plant tissues is

(a) 15 ppm   (b) 20 ppm   (c) 100 ppm   (d) 10 ppm

79. The average concentration of manganese in plant tissues is

(a) 15 ppm   (b) 20 ppm   (c) 100 ppm   (d) 10 ppm

80. The average concentration of zinc in plant tissues is

   (a) 15 ppm   (b) 20 ppm   (c) 100 ppm   (d) 10 ppm

81. The average concentration of copper in plant tissues is

   (a) 15 ppm   (b) 20 ppm   (c) 6 ppm   (d) 10 ppm

82. The average concentration of molybdenum in plant tissues is

   (a) 1.5 ppm   (b) 2.0 ppm   (c) 1.0 ppm   (d) 0.1 ppm

83. The average concentration of molybdenum in plant tissues is

   (a) 1.5 ppm   (b) 2.0 ppm   (c) 1.0 ppm   (d) 0.1 ppm

84. Mitschelich equation to relate potential yield of a crop to the amount of a given nutrient

   (a) $\log(A-Y) = \log A - Cb$ (b) $\log(Y) = \log A - Cb$

   (c) $\log(A-Y) = \log A + Cb$ (d) $\log(A-Y) = \log A / Cb$

85. According to Fick's first of diffusion

   (a) $F = -D/(dC/dx)$ (b) $F = -(D)(dC/dx)$

   (c) $F = +(D)(dC/dx)$ (d) $F = -(D)(dC*dx)$

86. The unit of diffusion coefficient in soil is

   (a) cm^2/s^2 (b) cm/s (c) cm^2/s (d) cm/s^2

87. The absorption of nutrients by plants increases with the concentration in the soil

   (a) linearly   (b) curvilinearly

   (c) hyperbolically   (d) parabolically

88. pH range of cell cytoplasm

   (a) 4.5-5.9   (b) 7.3-7.6   (c) 5.5-6.5   (d) 6.5-7.5

89. pH range of cell vacuole

   (a) 4.5-5.9   (b) 7.3-7.6   (c) 5.5-6.5   (d) 6.5-7.5

90. pH of cell apoplasm

   (a) 4.5-5.9   (b) 7.3-7.6   (c) 5.5   (d) 6.5

91. Excess of chlorides in soil results in lesser uptake of which element

   (a) Ca   (b) K   (c) Mo   (d) S

91. Mg is antagonistic to which of the following nutrient

   (a) Ca   (b) Mn   (c) P   (d) K

92. Excess of phosphates in soil results in lesser uptake of which element

   (a) Ca   (b) Mn   (c) P   (d) K

FILL IN THE BLANKS

1. _________ fraction of humus is insoluble in both acid and alkaline condition.

2. Little leaf of cotton is due to the deficiency of _________ .

3. _________ element is involved in the activity of CO_2 fixation enzymes.

4. Element present in Coenzyme A is _________ .

5. Exanthema disease of fruit trees is due the deficiency of _________ .

6. Mottling in citrus is due to deficiency of _________ .

7. Ideal concentration of CO_2 for plant growth is _________ ppm.

8. Fe^{3+} increase _________ fold with unit increase in pH.

9. Zn increase _________ fold with unit increase in pH.

10. Nutrient content is considered deficient when it limits _________ and produces _________ .

11. Nutrient content is considered _________ when it limits growth but not accompanied by the appearance of deficiency symptoms.

12. Nutrient element required in larger amounts by plants are termed as _________ .

13. Nutrient element required in moderate amounts by plants are termed as _________ .

14. Nutrient element required in smaller amounts by plants are termed as _________ .

15. The concentration range of Zn in matured plant leaf tissue to be considered as sufficient is _________ mg/kg.

16. The process of transformation of amines/ amino acids to ammonia by microbial enzymes is _________ .

17. The nitrogen percentage in atmosphere is _________ .

18. Best example for associative nitrogen fixing bacteria is _________ .

19. An element which is an essential constituent of proteins and are largely functional rather than structural is _________ .

20. The end products of denitrification process _________ and _________ .

21. Denitrification takes place at redox potential less than _________ mV.

22. List the catalyst and activator nutrient elements _________ .

23. Two forms of organic nitrogen _________ and _________ .

24. Nitrate limit in drinking water is _________ mg.

25. Plant mostly taken P as __________.

26. Inorganic forms of P are __________, __________ and __________.

27. P fixation takes place in _____________, _____________ and different forms.

28. Amorphous mineral that fix P is _______.

29. The constituent of nucleic acid ,phytin, phospholipids and phosphoprotein is _______.

30. Only mineral source of P in soil is __________.

31. Important K bearing minerals are __________ and __________.

32. K in soil is present in _________, __________ and _____________ different forms.

33. Q/I relationship given by __________ in _______.

34. The plant that show no obvious symptoms the nutrient content is sufficient is termed as _______.

35. The intake by a plant of an essential nutrient in amount exceeding what it needs is ___________.

36. Mycorrhiza help in the uptake of __________ and _______ nutrients.

37. Secondary nutrients which are immobile in plant are _______ and ___________.

38. Exhanthima disease is caused due to deficiency of _______ nutrient.

39. The major form of nitrogen in soil is _______.

40. Urea is readily soluble in water due to its _____________ nature.

41. The nutrient element essential for the transfer of genetic code to the off-springs is _______.

42. ___________ and _________ are the energy currency of plants.

43. ATP stores ______ kcal energy per mole.

44. In seeds, Ca is present in _________ form.

45. The nutrient element required for activation of phospholipase and ATPase enzymes is _______.

46. Upward curling of older leaves at margin is the deficiency symptom of ______ nutrient element.

47. The ___________ element is involved in the synthesis of glucosides in mustard.

48. The puckered with inward raised areas between the veins in older leaves is the deficiency symptom of ______ nutrient.

49. Iron exhibits two oxidation states because of its ________ nature.

50. ________ nutrient element is involved in the gibberellic acid metabolism.

51. ________ nutrient is essential to enhance fertility of male flowers.

52. _______ nutrient forms a part of plastocyanin enzyme and influences photosynthesis.

53. _______ nutrient is involved in transfer of electrons to N_2 and hence in biological nitrogen fixation required for nodulation in legumes.

54. __________ nutrient element transfers electrons to NO_3^- by nitrate reductase enzyme.

55. Boron is absorbed by the plants in the form of ________.

56. Heart rot of sugarbeet is due to deficiency of ________ nutrient.

57. Internal cork of apple is due to deficiency of ________ nutrient.

58. Top sickness of tobacco is due to deficiency of ________ nutrient.

59. _____ nutrient element is required for optimum hydrogenase activity and especially in free living N fixers.

60. Grey leaf spot in coconut palms can be managed by ________ element.

61. Supply f higher amount of sulphate can reduce the uptake of _____ions.

62. White muscle disease in livestock is the caused due to deficiency of _____.

63. Muscular dystrophy in livestock is the caused due to deficiency of _____.

64. According to Leibig's law of minimum "the nutrient that is present in least relative amount is the __________ nutrient".

65. According to Mitscherlich's physiological law "Yield can be increased by each single growth factor even when it is not present in the ________ as long as it is not present in ____________ "

66. According to Mitscherlich's growth law, "increase in yield of a crop as a result of increasing a single growth factor is proportional to ______ from the maximum yield obtainable by increasing that particular growth factor".

67. According to percentage yield concept of Baule, when one or more nutrient is deficient, the final percentage sufficiency is __________.

68. Law of maximum of Wallace and Wallace states that ________.

69. According to Bray's concept of nutrient availability, available forms of nutrients are thos forms whose variations in amounts in soil are responsible for significant variations in ____ and _____.

70. According to Bray's concept of nutrient mobility over all processes whereby nutrients reach ________, thereby making possible their sorption into plants.

71. Available nutrient is the one that is present in a pool of nutrients having an effective diffusion co-efficient larger than _________ cm^2/g.

72. The rate of nutrient transfer by mass flow is given by the rate of _________ and _____________.

73. Movement of nutrient ion in response to the concentration gradient is termed as __________.

74. Rate of chemical reaction doubles for every __ °C rise in temperature.

75. Application of gypsum to sodic soils leads to _______ in pH of soil and diffusion coefficient _______.

76. Theory of cation exchange was propounded by ____ and ________.

77. Root interception theory is the modified form of _____ theory.

78. The electrical potential difference of root membrane is ______ to ____ mV

79. Absorption of ions by plant roots along the electrochemical gradient is termed as __________.

80. Absorption of ions by plant roots against the electrochemical gradient is termed as __________.

81. Electrochemical theory of ion uptake was proposed by the scientist ______.

82. According to electrochemical theory of ion uptake _______ gradient existing between cell membrane and outside solution is responsible for migration of electrons from ________ to _______.

83. Electro osmotic theory of ion uptake was proposed by _______.

84. Plant cell membrane ATPase are located on ____ and ___ .

85. ATP ase ion pump theory was proposed by __________.

86. The ATPase –H^+ pumps work on _____ and ________ potential driving force.

87. __________ force is responsible for the transportation of protons against an electrochemical gradient across the cell membrane .

88. Passive transport of cations across plasmalemma into cytoplasm along electrical potential gradient is termed as ______

89. Carrier hypothesis of ion uptake was proposed by ________

90. According to Carrier hypothesis of ion uptake, ions transported to the surface of plant roots from soil combines with ______ molecules to form ______.

91. The process of conversion of organic form of nitrogen to inorganic form was first explained by the scientist ________.

92. The process of conversion of protein to aminoacids and amines by micro organisms is termed as __________.

93. The concept of P fixation in soil was first demonstrated by ________.

94. The nutrient element considered as master cation is ____________.

95. In quantity-intensity relationship of potassium, intensity factor is the measure of ____________ and quantity factor is the measure of ____________.

96. Formation of insoluble ________ is the cause of P induced Zn deficiency in soils and plants.

97. Nutreint index concept for soil fertility evaluation was first introduced by ___________.

98. Red colour in soil fertility map indicates ________ .

99. Soil test crop response approach was initiated by the scientist ______.

100. Fullform of DRIS is ___________.

101. DRIS is a system which identifies ________________ and helps in enhancing ____________ by improving __________.

102. DRIS system was first developed by the scientist ____________.

103. Establishment of ______ through standard methods is necessary in DRIS system.

104. A- value concept was first given by __________.

105. Pahala blight of sugarcane is the deficiency symptom of ___________ nutrient

106. White bud of maize is the deficiency symptom of ___________ nutrient.

107. Whip tail of cauliflower is the deficiency symptom of ______ nutrient.

108. Heart rot of sugarbeet is the deficiency symptom of ___________ nutrient.

109. Lime induced iron chlorosis is most common in ________ soils.

110. Ammonuim fertilizers are not recommended to ____ soils.

111. K fixation is absent in soils dominated with _________ clays.

112. General recommendation concentration of foliar spray of urea shouldnot exceed ______%

113. ________ was the first scientist to state that hummus is a source of food for plants.

114. The first agricultural research station in France was started by _________.

115. _________ and _______ elements are are classified as structural elements.

116. The fact that nitrogen compound are transformed soon to nitrates was first shown by _________.

117. The fact that non leguminous plants also fix nitrogen was first shown by _________.

118. Steenberg effect refers to _________ nutrients.

119. Mineral theory of plant nutrition was first given by the scientist _______.

120. The concept of cation exchange capacity of soil was first given by the scientist _______.

121. The method of DTPA extractable micro nutrients was first formulated by ______.

122. Deficiency of Fe, Zn and Mn can be observed in _________ soils

123. During nitrification process, ammonium is first converted to ___________ and then to _________ to get converted to nitrite.

124. On application of monocalcium phosphate to soil, it gets converted to ________.

125. ______ is the most important factor that regulates the solubility and availability of micronutrients in soil.

126. When the rate of dry matter production exceeds rate of nutrient uptake, the concentration of that nutrient in dry matter decreases, this is known as _________.

127. _________ form of N loss can be reduced by the use of nitrate fertilizers and deep placement.

TRUE OR FALSE

1. 'O' horizon is absent in arable soils.

2. Fulvic acid is soluble in alkali and insoluble in acid.

3. Magnesium is involved in rubisco enzyme activity.

4. Copper deficiency causes reclamation disease in cereals.

5. Crinkle leaf of cotton is due to Zn deficiency.

6. Bitter pit in apple is due to Mn deficiency.

7. Co is the 17th essential known element.

8. Nutrient element is one that is required to complete the life cycle of of the organism and its relative deficiency doesnot produce specific deficiency symptoms.

9. The sufficient range of K in matured leaf tissue of plant is 1- 5 %.

10. Urease enzyme is involved in conversion of amide form to ammonium carbonate in soil

11. The elements which are most deficient in Indian soils are N and K

12. The best example for symbiotic nitrogen fixing bacteria is rhizobium.

13. The element present in porphyrin structures of chlorophyll molecule is only Mg.

14. Higher hydrolysis of urea takes place in high pH.

15. The essentiality of nitrogen was given by Theo De Saussure.

16. K is considered as regulatory and carrier element.

17. Older leaves appear yellowish or chlorosis due to deficiency of K.

18. LCC is used to estimate N element

19. Nitrogen fixation is a mechanical process.

20. Phosphorus fixation is purely a chemical process.

21. P fixation is also called as reversion.

22. Montmorllionite mineral fixes highest amount P in soil.

23. P fixation decreases if increase in the concentration of organic matter in soil.

24. Deficiency of P leads to purple colouration on midrib of leaves.

25. K is the element which control the ionic balance and osmotic regulation.

26. Deficiency of K leads to the marginal necrosis of younger leaves.

27. Ca element is essential for sugars translocation in sugars cane.

28. Zn nutrient imparts winter hardiness to plant.

29. The element Ca involve directly in the chromosome stability and constituent of chromosome structure.

30. Grass tetany disease is due to deficiency of Mg.

31. Cu is essential for biosynthesis of tryptophan.

32. P nutrient is taken up by plant as combined anion.

33. The major forms of nitrogen in soil is inorganic form.

34. Nitrapyrin is an example for nitrification inbibitor.

35. Dicyandiamide is an example for slow releasing N fertilizer.

36. Oxamide is an example for nitrification inbibitor.

37. Potassium azide is an example for slow releasing N fertilizer.

38. N serve acts as both nitrification and urease inhibitor.

39. Tetrazole acts as both nitrification and urease inhibitor.

40. Under the conditions of large availability of N, succulence of plants increases.

41. Nitrogen rich, luxuriant succulent crop is resistant to pest and diseases.

42. Under reduced K supply, transportation of amino acids will be reduced.

43. Potassium is required for the maintenance of cytoplasmic pH.

44. Potassium is required for the activation and synthesis of nitrate reductase enzyme.

45. Calcium is required for the maintenance of chromosome structure.

46. Grass tetany is a nutritional disorder in cattle grazing on Mg-deficient pastured well fertilized with potassium.

47. Sulphur is a constituent of ferrodoxin and plays a role in biological nitrogen fixation.

48. The crop plants with N:S ratio more than 16:1 are suspected to be deficient in mustard.

49. Sulphur is mobile in plants hence its deficiency is seen first on younger leaves.

50. Manganese plays an important role in detoxification of superoxide free radicals.

51. Zinc is involved in the synthesis of RNA.

52. Under excessive Zn condition, excessive translocation of P occurs.

53. Rossetting of intenodes in dicots is the deficiency symptom of zinc.

54. Little leaves in dicots is the deficiency symptom of zinc.

55. Frenching in citrus is the deficiency symptom of zinc.

56. Molybdenum is the constituent of nitrogenase enzyme.

57. Molybdenum affects formation and viability of pollens and development of anthers.

58. Boron is not a constituent of any enzyme.

59. Boron is responsible for formation of cell wall and its stabilization.

60. Decrease in IAA can also be responsible for the induction of calcium deficiency.

61. Nickel influences urease activity and hence associated with nitrogen metabolism.

62. Chlorine plays a major role in osmoregulation and charge compensation in higher plants.

63. Take all disese in wheat can be managed by copper.

64. Chlorine lowers NO3- concentration in tissues and improves nutritional quality of vegetables.

65. Chlorine improves water balance of plants under limited water supply.

66. Silicon improves rigidity and strengthens cell wall.

67. Silicon aggravates P toxicity in plants.

68. Cobalt improves nutritional quality of the forage crops for ruminants.

69. Sulphate and selenate compete for the common uptake sites on roots.

70. Aluminium alleviates P toxicity in plants.

71. Aluminium alleviates Cu toxicity in plants.

72. Aluminium aggravates Zn toxicity in plants.

73. Vanadium favours nitrogen fixation in legumes.

74. If all the essential elements present except one, absence of that one constituent doesnot affect crop.

75. No_3^- is mobile in soil.

76. NH_4^+ is mobile in soil.

77. Available nutrient is the one that is present in the easily absorbed form by plant.

78. The rate of nutrient transfer by mass flow is indirectly related to the concentration of ion in the solution.

79. In dry soil, mass flow of nutrients doesnot occurs.

80. Low temperature enhances evaporation and transpiration.

81. Under concentration gradient condition, ions move from point of lower concentration to higher.

82. The higher the moisture content, higher will be the diffusion coefficient always.

83. For diffusion to occur, continuity of moisture films is a must.

84. With increase in soil compaction, the diffusion coefficient also increases.

85. With increase in soil temperature, the diffusion coefficient of nutrient ion also increases.

86. With increase in soil temperature, viscosity of water increases and thereby the diffusion coefficient of nutrient ion also increases.

87. Amelioration of acid soil leads to decrease in concentration of cations and vice versa for anions.

88. On application of organic manures to soil, the diffusion coefficient of nutrient ion also increases.

89. The magnitude of nutrients absorbed by plants depends only on the ability of soils to supply nutrients.

90. Electrochemical potential of ion in soil is directly related to the soil temperature.

91. Electrochemical potential of ion in soil is not related to the universal gas constant.

92. Root membranes of plant hold higher concentration of cations in cytoplasm than the outside concentration.

93. Root membranes of plant hold higher concentration of anions in cytoplasm than the outside concentration.

94. Cations are normally absorbed by plants in response to electrochemical potential gradient.

95. Anions are normally absorbed by plants in response to electrochemical potential gradient.

96. Active absorption of ions by plant roots takes place against the electrochemical gradient without the expenditure of metabolic energy.

97. Operation of ATPase –run H^+ pump is responsible for the negative charge of cell anad alkaline reaction of cytoplasm.

98. Toxicity of nitrous oxide to plants is common under alkaline conditions.

99. Nitrate is the form nitrogen majorly prone to leaching losses.

100. Higher the water table depth, losses of nitrogen will be higher.

101. Application of urea to acid soils reduces nitrogen losses.

102. Nitrogen losses are less under reduced condition than under aerated condition.

103. Newly formed alluvial soil contains low organic phosphorus.

104. Fe and Al- P in acid soil are insoluble form of soil phosphorus.

105. Ca-P in neutral soil is partially available form of soil phosphorus.

106. Hydroxy apatite in alkaline and calcareous soil are highly insoluble form of soil phosphorus.

107. Amorphus iron and aluminium oxides have larger P fixing capacity than crystalline silicates.

108. Kaolinite clays have larger P fixing capacity than monmorllionits clays.

109. Allophanes have larger P fixing capacity.

110. Calcareous soils have larger P fixing capacity.

111. Application of organic matter to soil reduces P fixation.

112. In general, fine textured soils are more capable of releasing potassium than coarse textured soil.

113. Acid soils reduce K fixation.

114. Zinc is less mobile in plants than P, therefore Zn accumulated largely in plant tissues than P.

115. Targeted yield equation obtained from soil test crop response approach can be extrapolated to other zones without any modifications.

116. Aluminium toxicity is observed in highly acidic soils.

117. The uptake of calcium is half that of N or K.

118. In general, finger millet cropping system mines more nutrients than maize cropping system.

119. The final product during symbiotic N fixation is glutamic acid.

120. P fixation is also referred to as P reversion.

121. Solubility of calcium phosphate is very low in high pH soils.

122. The longevity of residual P in soil depends on solubility of P fertilizers.

123. Lattice hole theory explains only the fixation of nitrogen nutrient.

124. On application of fresh organic residues, nitrogen locking occurs in soil.

125. The value of phosphatic fertilizers depend on initial reaction products.

126. P retention decreases with monovalent cation in the electrolyte.

127. Black soils are comparatively more fertile than laterite and red soils.

128. The important constituent of pyrrolr ring of chlorophyll is manganese.

129. Potassium is the only element classified as regulators and carriers.

130. Nitrogen and potassium are taken up by plants in almost equal proportion.

131. Organic manure application to crop fields on the day of sowing is recommended.

132. Flooding is the best way to manage acid sulphate soils.

133. Organic matter is considered as life of soil.

134. Amino acid is an example for hydrolysable N.

135. If the relationship between crop response and soil ttest values is curvilinear, then the logarithm of the data needs to be considered to obtain a linear scale.

136. The absorption pattern of different nutrients by palnts at their growth stags decides the time of fertilizer application.

137. Fertilizer recommendation should be based on cropping plan and availability of inputs.

MATCH THE FOLLOWING

1.

Scientists	Concepts
1. Justus Von Leibig	a. Mg
2. Dockuchaev	b. Law of diminishing marginal returns
3. Mitscherlich	c. Water is essential for plant growth
4. Helmont	d. Father of Agricultural chemistry
5. Glauber	e. Soil is a natural body
	f. Salt petre is a nutritional source

2.

Scientists	Concepts
1. Nicholas	a. Si
2. Wallerius	b. Father of plant analysis
3. Robert Bayle	c. Humus is the principle of vegetation
4. Arthur Young	d. Plants purify air
5. Joseph Priestly	e. Pot culture experiment
	f. Field experiment

3.

Term	Meaning
1. Sufficient nutrient	a. metal with specific gravity > 5 or having atomic number >20
2. Insufficient nutrient	b. Isotope of an element used for tracing its path in a system to study the mechanism of its interaction with the system
3. Deficient nutrient	c. found in low concentrations
4. Available nutrient	d. plays a role in plant metabolism, whether or not that role is specific or indispensable
5. Beneficial element	e. that fraction whose variation in amount is responsible for significant changes in yield and response
6. Nutrient element	f. stimulate plant growth but are not essential
7. Functional element	g. nutrient content in plants associated with optimum crop yields
8. Trace element	h. nutrient content in plants associated with only growth reductions but not appearance of deficiency symptoms
9. Tracer element	i. nutrient content in plants associated with growth reductions and appearance of deficiency symptoms
10. Heavy metal	j. required to complete the life-cycle of the organism

4.

Scientists	Relation
1. Nicholas	a. N
2. Aron and stout	b. Functional nutrient
3. Priestly	c. Essential nutrient
4. Theodore de Sausssure	d. Ca
5. Sprengel	e. CO_2

5.

Nutrient	Essentiality
1. K	a. Aron and Stout
2. S	b. Warrington
3. Fe	c. Sprengel
4. B	d. Sachs and Knop
5. Mo	e. Gris

6.

Nutrient	Sufficient or normal range in plants
1. N	a. 100-500 mg/kg
2. P	b. 0.1-0.4 mg/kg
3. Ca	c. 0.2-1.0 %
4. S	d. 1-5 %
5. Fe	e. 0.1-0.4 %

7.

Nutrient	Functions
1. N	a. Translocation of photosynthates
2. P	b . Translocation of carbohydrates
3. K	c. Improves quality of crop plants
4. Ca	d. Constituent of chlorophyll
5. Mg	e. Constituent of Co enzyme A
6. S	f. Constituent of volatile compounds
	g. Constituent of porphyrin molecule

8.

Nutrient	Functions
1. Fe	a. Constituent of plastocyanin protein
2. Mn	b . Pollen formation
3. Cu	c. Constituent of nitrate reductase
4. Zn	d. Constituent of porphyrin molecule
5. Ni	e. Constituent of urease enzyme
6. B	f. Involve in oxidative reactions
	g. Synthesis of tryptophan

9.

Nutrient	Deficiencies
1. N	a. Clorosis on edges of lower leaves
2. P	b. Clorosis of upper leaves
3. K	c. Yellowing of lower leaves
4. Ca	d. Intervienal clorosis of lower leaves
5. Mg	e. Purple colouration on margins of lower leaves
6. S	f. Purple colouration of upper leaves
	g. Yellowing of upper leaves

10.

Nutrient	Deficiencies
1. Fe	a. Clorosis of younger leaves
2. Mn	b. Breakdown of internal tissue
3. Cu	c. Necrosis of younger leaves
4. Zn	d. Intervienal clorosis of younger leaves
5. Ni	e. Necrosis of older leaves
6. B	f. Intervienal clorosis of younger leaves and retarded growth
	g. Intervienal clorosis of older leaves

11.

Nutrient	Mineral source
1. P	a. Gypsum
2. K	b. Dolomite
3. Ca	c. Apatite
4. Mg	d. Carbamate
5. C	e. Mica
	f. Tourmaline

12.

Nutrient	Mineral source
1.Mo	a. Chalcopyrite
2. Mn	b. Tourmaline
3. Cu	c. Necrosis of younger leaves
4. Zn	d. Olivine
5. Cl	e. Apatite
6. B	f. Pyrolusite
	g. Hornblende

13.

Biological method of soil evaluation	Nutrient
1. Sackett and Stewart technique	a. K
2. Mehlich technique	b. Cu and Mg
3. Mulder's technique	c. B
4. Mitcherlich pot culture	d. P
5. Sunflower pot culture	e. Ca
	f. N,P,K

14.

Nutrient	Indicator plant
1. N	a. Tobacco
2. P	b. Cauliflower
3. K	c. Maize
4. Ca	d. Potato
5. Mg	e. Tomato
6. Na	f. Coconut
	g. Sugarbeet

15.

Nutrient	Indicator plant
1. Fe	a. Citrus
2. Mn	b. Cauliflower
3. B	c. Tea
4. Cu	d. Potato
5. Mo	e. Sunflower
6. S	f. Cabbage
	g. Sugarbeet

16.

Nutrient	Antagonistic element
1. N	a. Mg
2. P	b. P
3. K	c. Mo
4. Ca	d. K
5. Mg	e. Zn
6. S	f. Mo
	g. S

17.

Nutrient	Antagonistic element
1. Fe	a. Al
2. Mn	b. Ca
3. Cu	c. N
4. Zn	d. Mo
5. Al	e. Cu
6. B	f. S
	g. P

18.

Enzymes	Functions
1. Carbonic anhydrase	a. Anaerobic respiration
2. Alcoholic dehydrogenase	b. Component of electron system of PS II
3. Superoxide dismutase	c. Controls senscence
4. Plastocyanin	d. Protection against oxidation
5. Polyphenol oxidase	e. Transfer of CO_2/HCO_3^- for photosynthetic CO_2 fixation
	f. Involved in lignin biosynthesis

KEY TO CHOOSE THE CORRECT ANSWER

1. a	2. b	3. d	4. a	5. b	6. a
7. d	8. b	9. c	10. c	11. d	12. a
13. d	14. d	15. c	16. d	17. b	18. b
19. d	20. c	21. b	22. c	23. c	24. c
25. d	26. d	27. c	28. b	29. d	30. c
31. a	32. d	33. b	34. c	35. d	36. a
37. a	38. k	39. d	40. c	41. c	42. a
43. b	44. b	45. b	46. c	47. a	48. b
49. d	50. a	51. d	52. b	53. a	54. d
55. b	56. d	57. a	58. b	59. b	60. c
61. a	62. d	63. b	64. b	65. c	66. d
67. d	68. a	69. a	70. b	71. b	72. b
73. c	74. a	75. b	76. c	77. c	78. b
79. b	80. b	81. c	82. d	83. d	84. a
85. b	86. c	87. c	88. b	89. a	90. c
91. c	92. c				

KEY TO FILL IN THE BLANKS

1. Humin	2. Zn
3. Mg	4. S
5. Cu	6. Zn
7. 1000	8. 100
9. 1000	10. Growth, characteristic deficiency symptoms
11. insufficient	12. Primary nutrients
13. Secondary nutrients	14. Micro nutrients
15. 27-150	16. ammonification
17. 78%	18. *Azospirillum*
19. N	20. N_2O, N_2

21.	+250	22.	Fe, Mn, Zn, Cu, B,Mo,Cl
23.	hydrolysable and non hydrolysable	24.	10mg
25.	$H_2PO_4^-$ Primary orthophosphate	26.	Fe-P,Al-P and Ca-P
27.	Adsorption, Isomorphic substitution and decomposition	28.	allophone
29.	phosphorus	30.	apatite
31.	mica, K feldspar	32.	non-exchangeable, exchangeable, solution K
33.	Beckette, 1960	34.	hidden hunger
35.	luxury consumption	36.	phosphorus and zinc
37.	calcium, sulphur	38.	copper
39.	organic form	40.	Non polar nature
41.	Nitrogen	42.	ATP and ADP
43.	12	44.	Ca-pectate
45.	Ca	46.	Mg
47.	Sulphur	48.	Sulphur
49.	Transitional	50.	Zinc
51.	Copper	52.	Copper
53.	Molybdenum	54.	Molybdenum
55.	H_3BO_3	56.	Boron
57.	Boron	58.	Boron
59.	Nickel	60.	Chlorine
61.	Selenate	62.	Selenium
63.	Selenium	64.	limiting
65.	minimum, optimum	66.	decrement

67. product of individual sufficiency

68. when the need is fully satisfied for every factor required in the process, the rate of the process can be at its maximum potential, which is greater than the sum of the individual parts because of the sequentially additive interactions

69.	yield and yield response	70.	sorbing root surfaces
71.	10^{-12}	72.	water flux into the root and the concentration of the ion in the solution

73. Diffusion
74. 10
75. Decrease, increases
76. Jenny and Overstreet
77. Theory of cation exchange
78. -120 to -180
79. Passive Absorption
80. Active Absorption
81. Ludenbargh
82. electrical potential, cytoplasm, apoplasm
83. Mittchell
84. Plasma membrane, tonoplast
85. Mengel and Kirkby
86. electrical and chemical
87. proton motive force
88. Uniport
89. Epstein
90. carrier, ion-carrier complex
91. Russel
92. Aminisation
93. Thomas Way
94. Potassium
95. potassium in soil solution, capacity of soil to maintain K in soil solution
96. Zinc phosphate
97. Parker
98. low
99. Ramamurthy
100. Diagnosis and Recommendation Integrated System
101. all the nutritional factors limiting crop production, crop yields, fertilizer recommendations
102. Beaufils
103. Calibrated norms
104. Fried and Dean
105. Manganese
106. Zinc
107. Molybdenum
108. Boron
109. Calcareous
110. Alkaline
111. Kaolinite
112. 1.0
113. Wallerius
114. Bossingualt
115. Carbon, Hydrogen
116. Warrington
117. Winodagradsky
118. Deficient
119. Liebig
120. Thomas and Way
121. Lindsay and Norwell
122. Saline alkali
123. Hydroxyl amine, hyponitrite
124. Bruschite
125. Soil pH
126. Dilution effect
127. Denitrification

KEY TO TRUE OR FALSE

1. True
2. False (True: Fulvic acid is soluble in alkali and in acid
3. True
4. True
5. False (True: Crinkle leaf of cotton is due to Ca deficiency)
6. False (True: Bitter pit in apple is due to Ca deficiency)
7. False (True: Ni is the 17th essential known element)
8. False (True: Nutrient element is one that is required to complete the life cycle of of the organism and its relative deficiency produce specific deficiency symptoms)
9. True
10. True
11. False (True: The elements which are most deficient in Indian soils are Zn and B)
12. True
13. False (True: The element present in porphyrin structures of chlorophyll molecule is N and Mg)
14. True
15. True
16. True
17. False (True: Older leaves appear yellowish or chlorosis due to deficiency of N)
18. True
19. True
20. True
21. True
22. False (True: Kaolinite mineral fixes highest amount P in soil)
23. True
24. False (True: Deficiency of P leads to purple colouration on margins of leaves)
25. True
26. False (True: Deficiency of K leads to the marginal necrosis of older leaves)
27. False (True: P element is essential for sugars translocation in sugars cane)
28. False (True: K nutrient imparts winter hardiness to plant)
29. True
30. True
31. False (True: Zn is essential for biosynthesis of tryptophan)
32. True
33. False (True: The major forms of nitrogen in soil is organic form)
34. True
35. True
36. False (True: Oxamide is an example for slow releasing N fertilizer)
37. False (True: Potassium azide is an example for nitrification inhibitor)
38. True

39. False (True: Tetrazole acts as both nitrification inhibitor)
40. True
41. False (True: Nitrogen rich, luxuriant succulent crop is susceptible to pest and diseases).
42. True
43. True
44. True
45. True
46. True
47. True
48. True
49. False (True: Sulphur is immobile in plants hence its deficiency is seen first on younger leaves.)
50. True
51. True
52. False (True: Under Zn deficit condition, excessive translocation of P occurs.)
53. True
54. True
55. True
56. True
57. True
58. True
59. True
60. True
61. True
62. True
63. False (True: Take all disese in wheat can be managed by chlorine).
64. True
65. False (True: Sodium improves water balance of plants under limited water supply).
66. True
67. False (True: Silicon counteracts zinc deficiency induces P toxicity in plants)
68. True
69. True
70. True
71. True
72. False (True: Aluminium alleviates Zn toxicity in plants)
73. True
74. False (True: Even if all but one of the essential elements be present, the absence of that one constituent renders the crop barren)
75. True
76. False (True: NH_4^+ is immobile in soil)
77. True
78. False (True: The rate of nutrient transfer by mass flow is directly related to the concentration of ion in the solution)
79. True
80. False (True: Low temperature reduces evaporation and transpiration)
81. False (True: Under concentration gradient condition, ions move from point of higher concentration to lower)
82. False (True: The higher the moisture content, higher will be the diffusion coefficient until the moisture reaches saturation)
83. True

84. True

85. True

86. False (True: With increase in soil temperature, viscosity of water reduces and thereby the diffusion coefficient of nutrient ion also increases)

87. True

88. True

89. False (True: The magnitude of nutrients absorbed by plants depends only on the plant need for those nutrients and ability of soils to supply nutrients)

90. True

91. False (True: Electrochemical potential of ion in soil is directly related to the universal gas constant)

92. True

93. False (True: Root membranes of plant hold higher concentration of cations in cytoplasm than the outside concentration)

94. True

95. False (True: Anions are normally absorbed by plants against electrochemical potential gradient)

96. False (True: Active absorption of ions by plant roots takes place against the electrochemical gradient by the expenditure of metabolic energy)

97. True

98. True

99. True

100. True

101. False (True: Application of urea to acid soils leads to increase in nitrogen losses because, during hydrolysis consumption of H ions increase soil pH)

102. False (True: Nitrogen losses are more under reduced condition because of increase in pH of soil and also carbonate and bicarbonate influences volatilization)

103. True

104. True

105. True

106. True

107. True

108. True

109. True

110. True

111. True

112. True

113. True

114. True

115. False (True: Targeted yield equation obtained from soil test crop response approach cannot be extrapolated to other zones. It is site specific)

116. True

117. True

118. True

119. True

120. True

121. True

122. True

123. False (True: Lattice hole theory explains only the fixation of nitrogen and also potassium)

124. True
125. True
126. True
127. True
128. False (True: The important constituent of pyrrolr ring of chlorophyll is magnesium)
129. False (True: Potassium , calcium and magnesium are the elements classified as regulators and carriers)
130. True
131. False (True: Organic manure application to crop fields two or three weeks prior to sowing is recommended)
132. True
133. True
134. True
135. True
136. True
137. True

KEY TO MATCH THE FOLLOWING

1.	1-d	2-e	3-b	4-c	5-f
2.	1-a	2-c	3-b	4-e	5-d
3.	1-g	2-h	3-i	4-e	5-f
	6-j	7-d	8-c	9-b	10-a
4.	1-b	2-c	3-e	4-a	5-d
5.	1-c	2-d	3-e	4-b	5-a
6.	1-d	2-e	3-c	4-b	5-a
7.	1-c	2-e	3-a	4-b	5-d	6-f
8.	1-d	2-f	3-a	4-g	5-e	6-b
9.	1-c	2-e	3-a	4-b	5-d	6-f
10.	1-d	2-f	3-a	4-g	5-e	6-b
11.	1-c	2-e	3-a	4-b	5-d	
12.	1-d	2-f	3-a	4-g	5-e	6-b
13.	1-d	2-a	3-b	4-f	5-c	
14.	1-c	2-e	3-a	4-b	5-d	6-f
15.	1-b	2-d	3-e	4-a	5-f	6-c
16.	1-c	2-e	3-a	4-b	5-d	6-f
17.	1-d	2-f	3-a	4-g	5-e	6-b
18.	1-e	2-a	3-d	4-b	5-f	

4   MANURES AND FERTILIZERS

Hamsa N. and Suma M.M

CHOOSE THE CORRECT ANSWER

1. Most of the plants take N in the form of
   (a) Nitrate   (b) Amide   (c) Ammonia   (d) Cyanamide

2. In the nitrification process, conversion of ammonium to nitrite is brought about by
   (a) Nitrobacter   (b) Nitrosomonas
   (c) Pseudomonas   (d) Bacillus

3. In the nitrification process, conversion of nitrite to nitrate is brought about by
   (a) Nitrobacter   (b) Nitrosomonas
   (c) Pseudomonas   (d) Bacillus

4. Nitrification process is
   (a) Biological oxidation   (b) Biological reduction
   (c) Biochemical oxidation   (d) Chemical oxidation

5. In arc process of nitrogen synthesis the end product is
   (a) Anhydrous ammonia   (b) Dilute HNO_3
   (c) Con(c) HNO_3 (d) Calcium cyanamide

6. N content in anhydrous ammonia
   (a) 46%   (b) 20%   (c) 26%   (d) 82%

7. In Haber-Bosch process the N_2 and H_2 gas mixture ratio is
   (a) 1:3   (b) 3:1   (c) 3:3   (d) 1:1

8. The principal constituents in producer gas is
   (a) CO_2+N_2 (b) $CO+N_2$ (c) $CO+N$ (d) CO_2+N

9. The principal constituents in water gas
   (a) $CO+H_2$ (b) CO_2+H_2 (c) CO_2+H (d) $CO+H$

10. Ammonium sulphate chemically is
    (a) $(NH_3)_2SO_4$ (b) NH_4SO_4 (c) NH_3SO_4 (d) $(NH_4)_2\ SO_4$

11. N content in ammonium sulphate
    (a) 20.6-21%   (b) 26%   (c) 28%   (d) 25%

12. Sulphur content in ammonium sulphate
    (a) 12%   (b) 25%   (c) 24%   (d) 20%

13. Ammonium sulphate is type of fertilizer
    (a) acidic   (b) Neutral   (c) Basic   (d) None of these

14. N content in NH_4NO_3
    (a) 33-35%   (b) 25-26%   (c) 20.6-21%   (d) 44-46%

15. NH_4NO_3 is type of fertilizer
    (a) Acidic   (b) Neutral   (c) Basic   (d) None of these

16. In CAN fertilizer N content is
    (a) 21%   (b) 25%   (c) 24%   (d) 20%

17. CAN is type of fertilizer
    (a) Acidic   (b) Basic   (c) Neutral   (d) None of these

18. N content in NH_4Cl
    (a) 21%   (b) 20%   (c) 23%   (d) 25%

19. NH_4Cl is type of fertilizer
    (a) Acidic   (b) Basic   (c) Neutral   (d) None of these

20. Chemically urea is
    (a) $CO+N_2$ (b) NH_4NO_3 (c) $CO(NH_2)_2$ (d) $(NH_4)_2SO_4$

21. N form in urea
    (a) Nitrate   (b) Amide   (c) Ammonia   (d) Cyanamide

22. Urea is type of fertilizer
    (a) Neutral   (b) Acidic   (c) Basic   (d) None of these

23. Biuret content formed from urea heating above this temperature
    (a) 100^0C (b) 150^0C (c) 90^0C (d) 170^0C

24. The permissible maximum biuret content of urea produced in India is
    (a) 0.5%   (b) 1.0%   (c) 1.5%   (d) 2.0%

25. The enzyme involved in hydrolysis of urea
    (a) Nitrogenase   (b) Nitrate reductase   (c) Urease   (d) Hydrogenase

26. N content in $NaNO_3$
    (a) 21%   (b) 20%   (c) 16%   (d) 25%

27. Na content in $NaNO_3$
    (a) 21%   (b) 20%   (c) 26%   (d) 30%

28. NH_4NO_3 is type of fertilizer
   (a) Acidic   (b) Basic   (c) Neutral   (d) None of these

29. N content in urea
   (a) 33-35%   (b) 25-26%   (c) 20.6-21%   (d) 44-46%

30. Which fertilizer is basic fertilizer
   (a) Urea   (b) CAN
   (c) Sodium nitrate   (d) Ammonium sulphate

31. Which fertilizer material is partly soluble in water
   (a) Urea   (b) CAN
   (c) Sodium nitrate   (d) Ammonium sulphate

32. Which fertilizer is highly hygroscopic
   (a) Urea   (b) CAN
   (c) Sodium nitrate   (d) Ammonium sulphate

33. Which fertilizer contain sulphur
   (a) Urea   (b) CAN
   (c) Sodium nitrate   (d) Ammonium sulphate

34. Nitro-lime stone is prepared from
   (a) $CaCO_3$ and NH_4NO_3 (b) $Ca(OH)_2$ and NH_4NO_3
   (c) CaO and $(NH_4)_2 SO_4$ (d) None of these

35. Monocalcium phosphate is
   (a) $Ca_3(PO_4)_2$ (b) $Ca(H_2PO_4)_2$ (c) $CaHPO_4$ (d) None of these

36. Dicalcium phosphate is
   (a) $Ca_3(PO_4)_2$ (b) $Ca(H_2PO_4)_2$ (c) $CaHPO_4$ (d) None of these

37. General formula for apatite
   (a) $Ca_3(PO_4)_2$ (b) $Ca(H_2PO_4)_2$ (c) $CaHPO_4$ (d) None of these

38. Basic material for manufacturing of phosphatic fertilizer is
   (a) Langbeinite   (b) kainite   (c) Apatite   (d) Sylvinite

39. Form of P in rock phosphate
   (a) Monocalcium phosphate   (b) Dicalcium phosphate
   (c) Tricalcium phosphate   (d) None of these

40. P_2O_5 content in SSP
   (a) 12%   (b) 16%   (c) 32%   (d) 60%

41. The colour of SSP is
   (a) White   (b) Red   (c) Grey   (d) Yellow

42. Sulphur content in SSP
    (a) 12% (b) 25% (c) 24% (d) 20%

43. Calcium content in SSP
    (a) 19-22% (b) 25-26% (c) 20.6-21% (d) 44-46%

44. SSP is type of fertilizer
    (a) Acidic (b) Basic (c) Neutral (d) None of these

45. Conc. Super phosphate is prepared from
    (a) RP+ H_3PO_4 (b) RP+HCl (c) RP+HNO_3 (d) RP+ H_2SO_4

46. P_2O_5 content in Conc. Super phosphate
    (a) 33-35% (b) 25-26% (c) 16-20 % (d) 44-52%

47. Nitro phosphate is prepared from
    (a) RP+ H_3PO_4 (b) RP+HCl (c) RP+HNO_3 (d) RP+ H_2SO_4

48. DAP is prepared from
    (a) NH_3+ H_3PO_4 (b) NH_3+HCl (c) NH_3+HNO_3 (d) NH_3+ H_2SO_4

49. N content DAP
    (a) 18% (b) 46% (c) 24% (d) 12%

50. P_2O_5 content in DAP
    (a) 18% (b) 46% (c) 24% (d) 12%

51. Conversion factor from P to P_2O_5
    (a) 0.44 (b) 1.2 (c) 2.29 (d) 0.83

52. Conversion factor from P_2O_5 to P
    (a) 0.44 (b) 1.2 (c) 2.29 (d) 0.83

53. MAP is reaction in
    (a) Acidic (b) Basic (c) Neutral (d) None of these

54. DAP is reaction in
    (a) Acidic (b) Basic (c) Neutral (d) None of these

55. Ammonium phosphate nitrate is a type of fertilizer
    (a) Acidic (b) Basic (c) Neutral (d) None of these

56. Ammonium phosphate sulphate is a type of fertilizer
    (a) Acidic (b) Basic (c) Neutral (d) None of these

57. In production of ammonium poly phosphate condension of orthophosphoric acid is done at above this temperature
    (a) 100° C (b) 150° C (c) 90°C (d) 213°C

58. Conversion factor from K to K_2O
    (a) 0.44   (b) 1.2   (c) 2.29   (d) 0.83

59. Conversion factor from K_2O to K
    (a) 0.44   (b) 1.2   (c) 2.29   (d) 0.83

60. K content in wood ash
    (a) 1-2 %   (b) 4-12%   (c) 20-22%   (d) 30-40%

61. In wood ash K in the form of
    (a) K_2O (b) K_2CO_3 (c) KCl   (d) K_2SO_4

62. Wood ash contain about this much Kg of borax per ton of ash
    (a) 4-11   (b) 20-30   (c) 30-40   (d) 40-60

63. Following mineral source is used in production of MOP
    (a) Sylvinite   (b) Kainite   (c) Carnallite   (d) Langbeinite

64. Following mineral source is used in production of SOP
    (a) Sylvinite   (b) Kainite   (c) Carnallite   (d) Langbeinite

65. SSP is produced by the reaction of rock phosphate with
    (a) HNO_3 (b) H_3PO_4 (c) HCl   (d) H_2SO_4

66. Conc. Superphosphate is produced by the reaction of rock phosphate with
    (a) HNO_3 (b) H_3PO_4 (c) HCl   (d) H_2SO_4

67. The chief component of single superphosphate that supplies P to plants is
    (a) Gypsum   (b) free phosphoric acid
    (c) monocalcium phosphate   (d) rock phosphate

68. Which is the acid soluble phosphatic fertilizer
    (a) Super phosphate   (b) Dicalcium phosphate
    (c) Rock phosphate   (d) None of the above

69. The per cent content of N and P in DAP is
    (a) 20&20   (b) 18&46   (c) 46&18   (d) 16&20

70. Which is the water soluble phosphate
    (a) $Ca_3(PO_4)_2$ (b) $Ca(H_2PO_4)_2$ (c) $CaHPO_4$ (d) None of these

71. The fertilizer that contains maximum S is
    (a) Nitric phosphate   (b) con(c) super phosphate
    (c) Single super phosphate   (d) enriched superphosphate

72. The superphosphate is largely produced, now a days from
    (a) Phosphate guanos   (b) Bone meal
    (c) Rock phosphate   (d) Basic slag

73. The chemical formula of ammonium phosphate is

    (a) $NH_4H_2PO_4$

    (b) NH_4PO_3

    (c) $(NH_4)_3HP_2O_7$

    (d) $(NH_4)_4P_2O_7$

74. Which fertilizer is highly explosive

    (a) DAP

    (b) Ammonium Nitrate

    (c) Ammonium Sulphate Nitrate

    (d) Ammonium Sulphate

75. Which fertilizer supplies S

    (a) DAP
    (b) MOP
    (c) CAN
    (d) SSP

76. Which one of the following pairs of plant nutrients other than phosphorus
    are present is superphosphate

    (a) Ca and Fe
    (b) S and N
    (c) Ca and S
    (d) Ca and K

77. Chemical formula of Sylvinite

    (a) KCl.NaCl

    (b) KNO_3

    (c) $K_2SO_4, 2MgSO_4$

    (d) $KCl,MgCl_2.6H_2O$

78. Chemical formula of Langbeinite

    (a) KCl.NaCl

    (b) KNO_3

    (c) $K_2SO_4, 2MgSO_4$

    (d) $KCl,MgCl_2.6H_2O$

79. Muriatic acid is

    (a) HNO_3
    (b) H_3PO_4
    (c) HCl
    (d) H_2SO_4

80. The mineral that contains K_2SO_4

    (a) Sylvinite
    (b) Kainite
    (c) Carnallite
    (d) Langbeinite

81. The % K_2O in KCl is

    (a) 16
    (b) 60
    (c) 48
    (d) 12

82. The formula of potassium metaphasphate is

    (a) KH_2PO_4
    (b) KPO_3
    (c) K_2HPO_4
    (d) $K_4P_2O_7$

83. Which fertilizer is most imported in India

    (a) MOP
    (b) CAN
    (c) Urea
    (d) SSP

84. The chemical formula of MOP is

    (a) KCl
    (b) KPO_3
    (c) K_2SO_4
    (d) $K_4P_2O_7$

85. The chemical formula of SOP is

    (a) KCl
    (b) KPO_3
    (c) K_2SO_4
    (d) $K_4P_2O_7$

86. The organic chemical fertilizer is

    (a) Urea

    (b) Sodium Nitrate

    (c) CAN

    (d) Ammonium Sulphate

87. The SSP is manufactured during
   (a) 1900   (b) 1905   (c) 1901   (d) 1906

88. Destructive distillation of coal done at the temperature
   (a) 1000^0 C   (b) 1500^0 C   (c) 2000^0C   (d) 3000^0C

89. DAP is a
   (a) Straight fertilizer   (b) complex complete fertilizer
   (c) complex incomplete fertilizer   (d) organic fertilizer

90. K content in KCl
   (a) 16-20%   (b) 38-48%   (c) 50-52%   (d) 10-12%

91. MOP is a type of fertilizer
   (a) Acidic   (b) Basic   (c) Neutral   (d) None of these

92. SOP is a type of fertilizer
   (a) Acidic   (b) Neutral   (c) Basic   (d) None of these

93. Which is the secondary nutrient fertilizer
   (a) DAP   (b) Ammonium Nitrate
   (c) MOP   (d) Magnesium Sulphate

94. Which is the micronutrient fertilizer
   (a) DAP   (b) Ammonium Nitrate
   (c) Borax   (d) Magnesium Sulphate

95. The fertilizer which supply only one primary nutrient is termed as
   (a) Straight fertilizer   (b) complex complete fertilizer
   (c) complex incomplete fertilizer   (d) organic fertilizer

96. The fertilizer which supply two primary nutrient is termed as
   (a) Straight fertilizer   (b) complex complete fertilizer
   (c) complex incomplete fertilizer   (d) organic fertilizer

97. 15:15:15 is a example of
   (a) Straight fertilizer   (b) complex complete fertilizer
   (c) complex incomplete fertilizer   (d) organic fertilizer

98. The P fertilizer is suitable for acid soil
   (a) Super phosphate   (b) Dicalcium phosphate
   (c) Rock phosphate   (d) None of the above

99. Which is not a fertilizer
   (a) KCl   (b) NH_4Cl (c) NH_4NO_3 (d) P_2O_5

100. Fertiliser control order came into force from
   (a) 1957   (b) 1955   (c) 2006   (d) 1857

101. FCO (2006) is a set of legally enforceable executive order issued by the Govt. of India under
    (a) Essential commodities act 1945
    (b) Fertiliser control act 1957
    (c) Fertiliser control act 1955
    (d) Essential commodities act

102. Schedule II and part A of Fertiliser control order describes
    (a) Lists of tolerance limits in plant nutrients and physical parameters for various fertilizers
    (b) Specifications of all fertilizers that can be marketed in India
    (c) Procedure for withdrawal of fertilizer samples
    (d) All the above

103. The fertiliser association of India was set up in
    (a) 1957
    (b) 1955
    (c) 2005
    (d) 1857

104. The fertiliser obtained by physically mixing two or more fertilisers is
    (a) Mixed fertiliser
    (b) Straight fertiliser
    (c) Complex fertiliser
    (d) Compound fertiliser

105. Rice plants prefer nitrogen mostly in the form of
    (a) N gas
    (b) Nitrate
    (c) Amide
    (d) Ammonium

106. Naptha, a distillate obtained during fractional distillation of petroleum in temperature range of
    (a) 25-30 °C
    (b) 200-250°C
    (c) 300-450 °C
    (d) 2000-3000 °C

107. The amount S in soil contributed by rainfall per ha per year is
    (a) 30-300 kg
    (b) 30-300 ppm
    (c) 3-30 %
    (d) 3-30 kg

108. The order of hygroscopicity of fertilizers.
    (a) Ammonical fertilizers >K fertilizers>P fertilizers
    (b) Ammonical fertilizers <K fertilizers<P fertilizers
    (c) Ammonical fertilizers >K fertilizers<P fertilizers
    (d) Ammonical fertilizers >K fertilizers=P fertilizers

109. The order of hygroscopicity of fertilizers
    (a) Urea<Ammonium sulphate<CAN
    (b) Ammonium sulphate<CAN< Urea
    (c) Urea>Ammonium sulphate>CAN
    (d) Ammonium sulphate>CAN> Urea

110. The rate of decomposition is higher in ________ stage of composting
    (a) Mesophilic
    (b) Thermophilic
    (c) All the stages
    (d) Psychrophilic

FILL IN THE BLANKS

1. The materials that are organic in origin, bulky and concentrated in nature and supply plant nutrients and also improve soil physical environment are termed as __________

2. The manures that are organic in origin and supply plant nutrients in less amounts are termed as __________.

3. The practice of ploughing into soil the undecomposed green plant tissues for the purpose of improving soil physical, chemical and biological environment is termed as ________.

4. Fertiliser control order consists of ________ schedules.

5. Lists of tolerance limits in plant nutrients and physical parameters for various fertilizers is indicated in schedule ________ and part ____ of Fertiliser control order

6. The first fertilizer produced in India is ________ in __________.

7. A single compound containing one primary nutrient only is termed as ______.

8. Nitrogen in __________ form readily absorbed by most of the plants and ______ form is taken up by only few plants.

9. Arc process involves a reaction between N and O at a temperature of __________ °(C)

10. ____acid is the end product in phosphate fertiliser industry.

11. ________ refers to the guaranteed minimum percentage of N, P and K contained in the mixed fertilizer.

12. ______ refers to the ratios of percentages of N, P and K in the fertilizer mixture.

13. _____ is a weight makeup material added to fertilizer ingredients to produce a mixture of the desired grade.

14. Fertilizers containing ammonia should not be mixed with basically reactive fertilizers because ____________.

15. Water soluble phosphatic fertilizers should not be mixed with free lime containing materials because _________.

16. Lower values of critical relative humidity of fertilizer indicate ____ hygroscopicity.

17. On an average well decomposed FYM contains ______% N, ______% P and ______% K.

18 The combustible gas produced from Gobar gas compost plant is ______.

19. The composition of biogas is ________ % methane, ______% CO_2, _____% H_2 and ____ % N_2.

20. Gypsum is used as chemical preservation in FYM because it acts as ____.

21. Mesophic temperature range during composting is ________.

22. Thermophilic temperature range during composting is ________.

23. Well decomposed compost contains ________% of moisture.

24. Zinc heptahydrate contains __________ % Zn.

25. Copper pentahydrate contains __________ % Cu.

26. Manganese sulphate trihydrate contains __________ % Mn.

27. Borax contains __________ % (B)

28. Solubor contains __________ % (B)

29. The word Manure is derived from ________ word and that originated from _________ language

30. The process of increasing the productive capacity of land by adding plant foods to the soil in different forms is termed as _________.

31. Trench method of preparing FYM has been recommended by ________.

32. The main two components of FYM are ______ as solid portion and ________ as liquid portion.

33. During storage of manure, the urine and dung are decomposed and considerable amount of ______ is produced and is more prone to ________ losses.

34. The word compost is derived from the ______ word _________.

35. The process of converting organic matter in to manure in a short time by accelerating fermentation process under controlled conditions is called __________ .

36. Fullform of ADCO process is _______

37. ADCO process of composting was initiated by _________.

38. Activated compost method was developed by __________.

39. Indore method of composting process is developed in India by scientists ______ at the _________.

40. Optimum temperature required for vermicomposting is _________.

41. Optimum moisture required for vermicomposting is ________.

42. Optimum aeration required for vermicomposting is __________.

43. The gas produced during aerobic decomposition is ______.

44. The gas produced during anaerobic decomposition is ______.

45. The product obtained from night soil without any admixture of other organic waste materials is called __________.

TRUE OR FALSE

1. The manure of younger animals is richer in nutrients than that of older animals.

2. The manure from bullocks is richer in nutrients than that of milch cows.

3. Schedule I Part A of Fertiliser Control Order deals with specifications of all fertilizers that can be marketed in India.

4. Schedule II and part B of Fertiliser control order elaborate detailed methods of analysis of fertilizers.

5. The first manufacturer of SSP in India was EID-Parry India Ltd., Ranipet.

6. Time required for the completion of anaerobic method of decomposition is 5-6 months.

7. Fertiliser containing nitrate or ammonium are not readily soluble in water.

8. Synthesis of ammonia is endothermic reaction.

9. Conditioners are the materials added to fertilizer mixtures during their preparation for reducing hygroscopicity and improving their physical conditions.

10. Tobacco stems can be used as conditioner during manufacture of mixed fertilizers.

11. Dolomtic limestone can used as neutralizer material of fertilizer residual acididty.

12. Ash can be used as filler material in fertilizers.

13. Fertilizers containing ammonia should not be mixed with basically reactive fertilizers.

14. Easily soluble fertilizers can be used in preparation of mixed fertilizers and can be stored for long time before application.

15. Granulated fertilizers ensure easier and uniform absorption.

16. Higher values of critical relative humidity of fertilizer is desirable.

17. Urea is highly soluble fertilizer among the commonly used nitrogenous fertilizers.

18. Ammonium sulphate is slightly hygroscopic in nature.

19. Urea has acidic residual effect.

20. MOP has basic residual effect.

21. Loss of ammonia as gas is higher under high temperature.

22. Decrease in ammonia vapour pressure increases volatalisation of ammonia.

23. Potassium chloride contains 48 % of Cl.

24. SSP is the best example for water soluble P fertilizer.

25. Triple super phosphate is an example for water insoluble P fertilizer.

26. The method adopted for collection of dung, urine and litter primarily also decides the quality of manure.

27. Covered pit method is considered as the best method for storing manure.

28. Nutrients of manures are water soluble and these are liable to get washed by rain water .

29. Gypsum is the commonly used preservative for storage of manure.

30. SSP can also be used as preservative for storage of manure.

31. Loss of N in the form of ammonia is less from wet manure.

32. The bad odour produced from ammonia during decomposition can be overcome by the addition of gypsum.

33. Limestone can be used to accelerate the process of composting.

34. Time required for the preparation of compost takes 2-3 months.

35. The time period required for the completion of compost by urban composting method is 2-3 months.

36. Watering to vermicompost pits has to be stopped one week prior to harvesting.

37. Time required for the completion of aerobic method of decomposition is 3-4 months.

MATCH THE FOLLOWING

1.   **% N in N fertilizers**

N fertilizers	% N
1. Anhydrous ammonia	a. 22
2. Ammonium sulphate	b. 26
3. Urea	c. 46
4. CAN	d. 21
5. Ammonia solution	e. 80-82

2.

P fertilizers	% P
1. SSP	a. 14-18
2. DAP	b. 34-39
3. DSP	c. 16-18
4. Basic slag	d. 46
5. DCP	e. 32

3.

Fertilizer	Nutrient content
1. Single super phosphate	a. 17. 8 % S
2. Ammonium sulphate	b. 18 % S
3. Potassium sulphate	c. 15 % S
4. Gypsum	d. 24 % S
5. Zn suplphate	e. 9 % S

4.

Liming material	$CaCO_3$ equivalent (%)
1. Calcium oxide	a. 136
2. Calcium hydroxide	b. 109
3. Dolomite	c. 86
4. Calcite	d. 179
5. basic slag	e. 100

5.

Fertilizer	Equivalent acidity
1. Anhydrous ammonia	a. 80
2. Ammonium chloride	b. 77
3. Ammonium sulphate	c. 148
4. Urea	d. 128
5. DAP	e. 110

KEY TO CHOOSE THE CORRECT ANSWER

1. a	2. b	3. a	4. c	5. b	6. d
7. a	8. b	9. a	10. d	11. a	12. c
13. a	14. a	15. a	16. b	17. c	18. d
19. a	20. c	21. b	22. b	23. d	24. c
25. c	26. c	27. c	28. a	29. d	30. c
31. b	32. a	33. d	34. a	35. b	36. c
37. a	38. c	39. c	40. b	41. c	42. a
43. a	44. c	45. a	46. d	47. c	48. a
49. a	50. b	51. c	52. a	53. a	54. b
55. a	56. a	57. d	58. b	59. d	60. b
61. b	62. a	63. a	64. d	65. d	66. b
67. c	68. c	69. b	70. b	71. c	72. c
73. b	74. b	75. d	76. c	77. a	78. c
79. c	80. d	81. b	82. b	83. a	84. a
85. c	86. a	87. d	88. a	89. c	90. c
91. c	92. b	93. d	94. c	95. a	96. c
97. b	98. c	99. d	100. a	101. d	102. c
103. b	104. a	105. d	106. b	107. d	108. a
109. c	110. b				

KEY TO FILL IN THE BLANKS

1. Manures	2. Bulky organic manures
3. Green manuring	4. Two
5. I, B	6. SSP, 1906
7. Straight fertilizer	8. Nitrate, Ammonium
9. 3500	10. Phosphoric
11. Fertilizer grade	12. Fertilizer ratio
13. Filler	14. Loss of N occur in the form of gaseous nitrogen

15. Portion of soluble phosphates get converted into insoluble form
16. High
17. 0.5,0.2 and 0.5
18. Methane
19. 50-60, 30-45, 5-10 and 1-2
20. Ammonia-absorbing agent
21. 15-50 ºC
22. 45-65 ºC
23. 15-25
24. 21
25. 25
26. 26-28
27. 10.5
28. 19
29. Manoeuverer, French
30. Manuring
31. C.N.Acharya
32. dung, urine
33. ammonia, volatilisation
34. Latin, Componere
35. Composting
36. Agricultural Development Company
37. Hutchinson, H.B and Richards, E.H
38. Fowler and Ridge
39. Howard and Ward, Indian Institute of plant Industry, Indore
40. 18 – 35 ºC
41. 60-70 %
42. 50 %
43. carbon dioxide
44. methane
45. Poudrette

KEY TO TRUE OR FALSE

1. False (True: The manure of younger animals is poor in nutrients than that of older animals.)
2. True
3. True
4. True
5. True
6. True
7. False (True: Fertiliser containing nitrate or ammonium are readily soluble in water.)
8. False (True: Synthesis of ammonia is exothermic reaction)
9. True
10. True
11. True
12. True
13. True
14. False (True: Easily soluble fertilizers can be used in preparation of mixed fertilizers and cannot be stored for long time before application)

15. True
16. True
17. True
18. False (True: Ammonium sulphate is highly hygroscopic in nature)
19. True
20. False (True: MOP has acidic residual effect)
21. True
22. False (True: Decrease in ammonia vapour pressure decreases volatalisation of ammonia)
23. True
24. True
25. False (True: Triple super phosphate is an example for water soluble P fertilizer)
26. True
27. True
28. True
29. True
30. True
31. True
32. True
33. True
34. True
35. True
36. True
37. True

KEY TO MATCH THE FOLLOWING

1.	1-e	2-d	3-c	4-b	5-a
2.	1-c	2-d	3-e	4-a	5-b
3.	1-e	2-d	3-b	4-c	5-a
4.	1-d	2-a	3-b	4-e	5-c
5.	1-c	2-d	3-e	4-a	5-b

5 SOIL CHEMISTRY

Usha Kumari and J. Veena

CHOOSE THE CORRECT ANSWER

1. Who is the founder of soil chemistry?
   (a) J.B Boussingault
   (b) J.B Van Helmont
   (c) J.T Ways
   (d) Justus Von Liebig

2. The range of diameter of true colloids?
   (a) 2nm -200nm
   (b) Less than 2 nm
   (c) 3nm -300 nm
   (d) > 250nm

3. The size range particles of ionic solution ?
   (a) 2nm -200nm
   (b) Less than 2 nm
   (c) 3nm -300 nm
   (d) > 250nm

4. The typical property of colloidal solution?
   (a) Specific surface area
   (b) Homogeneous
   (c) < 1nm
   (d) All of the these

5. The most reactive components of the soil
   (a) Clay   (b) Silt   (c) Sand   (d) Clay + Silt

6. pH dependent charge is most dominant in _________
   (a) Clay fraction
   (b) Silt and Sand
   (c) Humus
   (d) All of the these

7. Permanent charges are due to
   (a) Protonation
   (b) Deprotonation
   (c) Isomorphic substitution
   (d) Both a & b

8. In Brucite mineral the positive charge develops due to
   (a) Replacement of Al by Mg
   (b) Replacement of Mg by Al
   (c) Replacement of Si by Al
   (d) pH variation

9. CEC is higher in minerals
   (a) 1:1 minerals   (b) 2:1 minerals   (c) 2:1:1   (d) Kaolinite

10. The pH at which the total positive charge is equal to total negative charge is known as
    (a) Zero point charge
    (b) Isoelectric point
    (c) Zeta potential
    (d) None of the these

11. Increasing the electro negativity of the metal, acidic strength is
    (a) Increases
    (b) Decreases
    (c) Not affected
    (d) Abruptly increases

12. Increasing the electro negativity of the metal, ZPC is
    (a) Increases
    (b) Decreases
    (c) Not affected
    (d) Abruptly increases

13. The pH at which ZPC attained for oxides of mineral
    (a) Si-2, MnO_2-4, Fe_2O_3 - 6.5-8.0
    (b) Si-4, MnO_2-4, Fe_2O_3 – 7.0-9.5
    (c) Si-2, MnO_2-2, Fe_2O_3 - 6.5-8.0
    (d) Si-6.6, MnO_2-4, Fe_2O_3 - 6.5-8.0

14. The development of charges on organic colloids is due to
    (a) Protonation
    (b) Deprotonation
    (c) Dissociation of surface functional groups
    (d) All of the these

15. Carboxylic group and phenolic group dissociates at
    (a) Higher pH and lower pH respectively
    (b) lower pH and Higher pH respectively
    (c) Both at higher pH
    (d) Both at lower pH

16. Organic matter rich soils may not have a zero point of charge as
    (a) No net positive charge develop on organic colloids
    (b) No net negative charge develop on organic colloids
    (c) Higher positive than negative charge is developed
    (d) None of the abov

17. Substances which is adsorbed on another is
    (a) Adsorbent
    (b) Adsorbate
    (c) Adsorption
    (d) Accumulator

18. AEC is more in mineral rich soil
    (a) Kaolinite rich soil
    (b) Montmorillonite
    (c) Chlorite
    (d) Vermiculite

19. The substance on which adsorption occurs
    (a) Adsorbent   (b) Adsorbate   (c) Adsorption (d) Accumulator

20. Fixed layer of charges in diffuse double layer is called as
    (a) Helmholtz layer   (b) Gouy-chipman layer
    (c) Stern layer   (d) None of the these

21. Clay particles are hydrophilic due to
    (a) Absorption   (b) Diffusion
    (c) Hydration   (d) None of the these

22. Clay particles are hydrophobic due to
    (a) Sensitivity to electrolytes   (b) Higher molecular weight
    (c) Presence of negative charge   (d) All of the these

23. Clay particles have higher specific surface area due to
    (a) Smaller size   (b) Plate like shape
    (c) Spherical shape   (d) Both a & b

24. Amorphous clays are commonly formed from
    (a) Secondary clay minerals   (b) Quartz
    (c) Both a & b   (d) Volcanic ashes

25. Acid soluble fraction humus is
    (a) Humic acid   (b) Fulvic acid
    (c) Humin   (d) Hymatomelanic acid

26. Most stabalized and resistant fraction of humus is
    (a) Humic acid   (b) Fulvic acid
    (c) Humin   (d) Hymatomelanic acid

27. CEC of soil affected by pH as
    (a) Increases with increase in pH
    (b) Decreases with decreases in pH
    (c) Not affected by pH
    (d) First increase than decrease with pH

28. Molecular weight is higher for humic acid fraction
    (a) Fulvic acid   (b) Humic acid
    (c) Humin   (d) None of the these

29. Optical density is more in
    (a) Fulvic acid   (b) humic acid
    (c) Humin   (d) None of the these

30. Which fractions of soil is considered as plasma?

    (a) Humus   (b) Clay

    (c) Sand and clay   (d) Both a &b

31. Plasma when deposited on the surface of skeleton grains is called

    (a) Cutan   (b) Duripan   (c) Both a & b (d) Pedon

32. Actual seat of soil chemicl activity ?

    (a) Clay   (b) Plasma

    (c) Organic plasma   (d) None of these

33. Redox potential is given by the equation:

    (a) $E_h = E^\circ + (RT/nF) \ln (\text{oxidation/ reduction})$

    (b) $E_h = E^\circ + (RT/nF) \ln (\text{reduction/ oxidation})$

    (c) $E_h = E^\circ + (RT/nF) \log (\text{reduction/ oxidation})$

    (d) None of these

34. Expansion of IUPAC is

    (a) International Union of Pure and Applied Chemistry

    (b) Indian Union of Pure and Applied Chemistry

    (c) Indian Union for Purification and Application of Chemistry

    (d) None of these

35. Soil with high electron activity is considered to be

    (a) Oxidised soil   (b) Reduced soil

    (c) Oxido-reduced   (d) All of these

36. The relationship between electron activity and electrode potential

    (a) $E_h = 0.59 \text{ Pe}$ (b) $E_h = 0.0059 \text{ Pe}$

    (c) $E_h = 0.059 \text{ Pe}$ (d) $E_h = 0.059 \text{Pe}$

37. Critical ODR value for soil is

    (a) $20 \times 10^{-6} \text{ g/cm}^2/\text{min}$ (b) $2 \times 10^{-10} \text{ g/cm/min}$

    (c) $0.2 \times 10^{8} \text{ g/cm/min}$ (d) $20 \times 10^{-8} \text{ g/cm}^2/\text{min}$

38. Dissociation constant of water

    (a) 1.01×10^{-14} (b) 1.8×10^{-16} (c) 1.8×10^{-12} (d) 1.01×10^{14}

39. Oxidation potential of a soil at equilibrium is given as

    (a) $E_h = E^\circ + (0.059/n) \log (\text{reduction/ oxidation})$

    (b) $E_h = E^\circ + (0.059/n) \log K_{eq}$

    (c) $E_h = E^\circ + (0.0059/n) \log K_{eq}$

    (d) $E_h = E^\circ + (0.059/nF) \log K_{eq}$

40. The value of Avogardo's number is
    (a) 6.013×10^{19}
    (b) 6.023×10^{21}
    (c) 6.023×10^{-23}
    (d) 6.023×10^{23}

41. The concentration of HPO_4^{2-} and $H_2PO_4^{-}$ ions in solution is equal at pH of
    (a) 5.6
    (b) 6.5
    (c) 7.2
    (d) 8.2

42. In the formation organ metallic complexes in soils, the metal ion acts as a/an.
    (a) Ligand
    (b) Electron donor
    (c) Electron acceptor
    (d) Inert ion

43. If the nitrate formation in a soil were Independent of the amount of ammoniacal-N, in that case the reaction would be considered as
    (a) Zero order
    (b) First order
    (c) Second order
    (d) Either first or second order

44. The molarity of concentrated H_2SO_4 (AR grade) is
    (a) 10 M
    (b) 12M
    (c) 16M
    (d) 18 M

45. Eh i.e. redox potential of calomel electrode (reference electrode) is
    (a) 2.43mV
    (b) 24.3mV
    (c) 243mV
    (d) 243V

46. Eh i.e. redox potential of a solution increases with
    (a) Increase in reduced sp.
    (b) Increase in oxidized sp.
    (c) Decrease in reduced species
    (d) Decrease in oxidized sp.

47. Thickness of DDL (Double Diffuse Layer) increases with
    (a) Increase in electrolyte concentration
    (b) Decrease valencyof an ion.
    (c) Increase in dielectric constant of medium
    (d) All a, b and c are true

48. Thickness of DDL (Double Diffuse Layer) decreases with
    (a) Decrease in electrolyte concentration
    (b) Increase valencyof an ion
    (c) Increase in dielectric constant of medium
    (d) All a, b and c are true

49. Divalent cations (Ca^{++}, Mg^{++}) _________ the thickness of DDL more strongly than that of monovalent ion (Na, K^+)
    (a) Increase
    (b) Decreases
    (c) Has no effect on
    (d) Firstly increase andthen decrease

50. pH of a soil-solution _________ with dilution which is called as dilution effect (According to DDL theory)

(a) Increases
(b) Decreases
(c) remains the same
(d) Firstly increases andthen decreases

51. pH of soil solution supernatant measured with glass and calomel a pair of electrode is found to be more/greater/higher than that of measured in soil sediment and this phenomenon is termed as

(a) Dilution effect
(b) Suspension effect
(c) H+ ion effect
(d) Both a and b

52. As the soil pH increases the negative charge on the soil colloid

(a) Increases
(b) Decreases
(c) Remains the same
(d) Firstly increases and then decreases

53. ZPC can be determined by

(a) Potentiometrically
(b) Gravimetrically
(c) Conductometrically
(d) Cannot be determined

54. At pH greater than ZPC soil colloids exhibit mostly

(a) Negative charge
(b) Positive charge
(c) Charge less/neutral
(d) both a and b

55. Which one is correct among following

(a) Activity= activity coefficient x concentration
(b) Activity= activity coefficient / concentration
(c) Activity x activity coefficient = concentration
(d) Activity / activity coefficient = concentration

56. Higher E4/E6 (7.1) ratio means dominance of (aliphatic nature)

(a) Humic acid
(b) Fulvic acid
(c) Humin
(d) None of these

57. Which of the following remain constant for an equilibrated ion in the soilsystem?

(a) Electrochemical potential
(b) Chemical potential
(c) Activity
(d) Electric potential

58. Ionic potential is maximum in case of

(a) Cs
(b) Li
(c) K
(d) Na

59. The pH scale was proposed by

    (a) Arnon   (b) Sorenson   (c) Liebig   (d) Munsel

60. The pH meter works on the basis of

    (a) Laplas equation   (b) Bohr's equation

    (c) Winstor equation   (d) Nernst equation

61. What is the pH of 0.001m HCl ?

    (a) 3.3   (b) 3.0   (c) 4.3   (d) 4.1

62. Range of redox potential at which NO^{3-} reduction occurs

    (a) – 220 to – 240 mV   (b) – 280 to – 220 mV

    (c) – 220 to – 260 mV   (d) – 260 to – 280 mV

63. Range of redox potential at which Mn reduction occurs

    (a) – 220 to – 240 mV   (b) – 280 to – 220 mV

    (c) – 220 to – 260 mV   (d) – 260 to – 280 mV

64. 1 mg of H^+ will be equal to

    (a) 20 mg of Ca^{2+} (b) 9 mg of Al^{3+}

    (c) 12 mg of Mg^{2+} (d) All

65. Persistent binding agents are due to

    (a) Microbial products   (b) Plant derived products

    (c) Fungal extension   (d) Humic materials

66. Concentration of pure water is

    (a) 55.5 mol/L   (b) 45.5 mol/L

    (c) 48.5 mol/L   (d) 57.5 mol/L

67. If all the octahedral positions are occupied by Mg^{2+} ions, it is called as

    (a) Tetrahedral subgroup   (b) Octahedral subgroup

    (c) Trioctahedral subgroup   (d) Dioctahedral subgroup

68. If all the octahedral positions are occupied by Al^{3+} ions, it is called as

    (a) Tetrahedral subgroup   (b) Octahedral subgroup

    (c) Trioctahedral subgroup   (d) Dioctahedral subgroup

69. Which ion contributed maximum acidity ion exchange complex

    (a) H^+ (b) Al^{3+} (c) Fe^{3+} (d) Mn^{2+}

70. What would be the pH of solution of H^+ ion concentration 3.2×10^{-4} moles/ L

    (a) 3   (b) 3.5   (c) 4   (d) 4.5

71. Transition of an electron from M level to K level results in
    (a) Absorption spectra
    (b) Emission spectra
    (c) X-ray spectra
    (d) Continuous spectra

72. Negative ions are formed from neutral atoms by gain of
    (a) Positrons
    (b) Neutrons
    (c) Protons
    (d) Electrons

73. An oxygen atom contains
    (a) 8 neutrons and 8 protons
    (b) 8 protons two neutrons
    (c) 8 protons and 16 electrons
    (d) 8 protons and 8 electrons

74. The pH of a solution is defined by the expression
    (a) $\log [H^+]$
    (b) $-\log 1/[H^+]$
    (c) $\log 1/[H^+]$
    (d) $1/\log [H^+]$

75. Which of the following is correct?
    (a) $pH + pOH = 14$
    (b) $pH \times pOH = 14$
    (c) $pH - pOH = 14$
    (d) $pH/pOH = 14$

76. The most electronegative element in the periodic table is
    (a) Chlorine
    (b) Sodium
    (c) Cesium
    (d) Fluorine

77. Compared to the first ionization potential of an atom the second is
    (a) The same
    (b) Greater
    (c) Smaller
    (d) Negligible

78. Variable valence is shown by
    (a) Metallic elements
    (b) Normal elements
    (c) Transitional elements
    (d) Non-metallic elements

79. The migration of colloidal piratical particles under the influence of electric filed is known as
    (a) Electrophoresis
    (b) Dialysis
    (c) Electroosmosis
    (d) Electrodialysis

80. Scattering of light by colloidal particles is called
    (a) Brownian movement
    (b) Tyndall effect
    (c) Electrophoresis
    (d) Electroosmosis

81. Increase in zeta potential of soil particles leads to
    (a) Flocculation
    (b) Deflocculation
    (c) Aggregation
    (d) Both a and b

82. What is the SI unit of cation exchange capacity?
    (a) meq/100g
    (b) cmol(p+) kg
    (c) mmho/100g
    (d) All

83. In per cent base saturation, [BS (%) = S/T × 100], S denotes
    (a) Total cations
    (b) Basic cations
    (c) Total anions
    (d) Total cations and anions

84. Examples of natural chelates are
   (a) Citric acid and oxalic acid
   (b) EDTA and DTPA
   (c) Citric acid, oxalic acid and HEDTA
   (d) EDTA, DTPA and HEDTA

85. Which of the following is/are (an) example (s) for reducing sugars?
   (a) Glucose
   (b) Fructose
   (c) Lactose
   (d) All

86. The electrode of the pH meter consists of what element in contact with a solution of potassium chloride?
   (a) Hg
   (b) Mg
   (c) Zn
   (d) Ag

87. Which of the following is a conversion factor for ppm and percentage?
   (a) $\% \times 10,000 = ppm/100$
   (b) $\% \times 10,000 = ppm/200$
   (c) $\% \times 10,000 = ppm$
   (d) None of these

88. The book 'Principle of Soil Chemistry' was written by
   (a) K.H. Tan
   (b) Kononova
   (c) Adams
   (d) Russel

89. The cation exchange capacity of soil containing predominantly kaolintic clay mineral generally range from
   (a) 3 – 15
   (b) 15 – 20
   (c) 80 – 100
   (d) 100 – 150

90. Increase in zeta potential of soil particles leads to
   (a) Flocculation
   (b) Deflocculation
   (c) Aggregation
   (d) Both a and b

91. Ionic radius of silicon
   (a) 1.32 Af
   (b) 0.42 Af
   (c) 1.58 Af
   (d) 1.18 Af

92. Due to submergence
   (a) The pH of acid soil is increased
   (b) The pH of alkali soil decreased
   (c) The pH of acid soil is decreased
   (d) Both a and b

93. What is the optimum pH for the availability of most of the plant nutrients?
   (a) 5.0 – 6.0
   (b) 6.0 – 7.0
   (c) 6.5 – 7.5
   (d) 6.0 – 8.0

94. Number of equivalents of a substances dissolved in 1L of a solution is called as
   (a) Molarity
   (b) Molality
   (c) Normality
   (d) Formality

95. A measured of the effective concentration of a reactant or product in a chemical reaction is called as
   (a) Activity
   (b) Activity co-efficient
   (c) Ionic strength
   (d) None

96. Ion activity co-efficient is calculate by
    (a) Donan equation
    (b) Debye –Huckel equation
    (c) Henderson –Hasselbalch equation
    (d) Van't Hoff equation

97. Equivalent weight of $K_2Cr_2O_7$
    (a) $1/12^{th}$ of molecular weight
    (b) $1/6^{th}$ of molecular weight
    (c) $1/4^{th}$ of molecular weight
    (d) $1/3^{rd}$ of molecular weight

98. The redox potential at which SO_4 is reduced in the waterlogged soil is
    (a) 100 mV
    (b) 200 mV
    (c) -100 mV
    (d) -200mV

99. Methanogenesis in waterlogged soil takes place at Eh of
    (a) <-200mV
    (b) 200-300mV
    (c) 300-400mV
    (d) >400mV

100. When phenolphthalein is added to an acid solution the solution becomes
    (a) Yellow
    (b) Colourless
    (c) Blue
    (d) Pink

101. In flocculation zeta potential
    (a) Increase
    (b) Decrease
    (c) Does nor change
    (d) Increase and then decrease

102. If H^+ ion concentration of a solution is 10^{-6} M, what will be the pOH?
    (a) 6
    (b) 8
    (c) 10
    (d) 14

103. The major source of negative charge on humus colloids is due to
    (a) Carboxylic groups
    (b) Enolic groups
    (c) Phenolic groups
    (d) All of these

FILL IN THE BLANKS

1. Write down unit of AEC (Anion Exchange Capacity) __________.

2. If the reaction is spontaneous when ÄG is __________.

3. If a solution has pH 4 and it is diluted 100 times then what would be the pH of solution ___________.

4. Individual crystalline particle is known as ____________.

5. If two solution having pH value of 5.0 and 6.0 are mixed in equal quantities, what shall be pH of resulting mixture ____________.

6. The pH of 0.1 N NaOH solution would theoretically be __________.

7. A solution of pH 4.0 is diluted with distilled water 100 times the pH of diluted solution is ________.

8. Ligand adsorption of P by soil results in ____________.

9. Thermodynamically a chemical system issaid to attain equilibrium at _________________________ .

10. Size of K ion is similar to that of _______________ .

11. The potential difference between the fixed part and freely mobile portion of the diffuse double layer is known as _________________ .

12. High zeta potential of a clay colloidal system represents the state of _______________ .

13. Acid and alkali insoluble fraction of humic substances _______________ .

14. The colloidal surface area in the upper 15cm of a hectare of clay soil be as high as Km_________ .

15. Number of equivalents of a substances dissolved in 1L of a solution is called as _____________ .

16. C/N/P/S ratio in sol humus is about _________________ .

17. The regular arrangement of points in space is called as_________________ .

18. _____________ has given the statement that the cation exchange reaction is the second most important phenomenon after photosynthesis.

19. Phosphorus fixation will____________ with increase in pH in the calcareous soil.

20. Ability of the system to resist drop in the redox potential is called as _____________ .

MATCH THE FOLLOWING

1.

	A			B
1	Ratio law		a	Variable charge surface
2	Bragg's law		b	Selectivity coefficient
3	ZPC		c	Donnan membrane equilibrium
4	Osmo-regulation		d	X-ray diffraction
5	Gapon equation		e	Salt-tolerance of crops

2.

	A (Reagents)		B (pH)	
1	Trough reagent		a	4.8
2	Olsen's reagent		b	3.0
3	Morgan's reagent		c	8.5

3.

	A (Reagents)			B (pH)
1	Ligand adsorption		a	Solute movement
2	Stoke's law		b	Ion exchange
3	Fick's law		c	Heavy metal
4	Gapon's equation		d	Settling velocity

4.

	A (Soil pH)			B (Predominant aluminium species)
1	< 4.7		a	$Al(OH)^{2+}$
2	4.7-6.5		b	Al^{3+}
3	6.5-8.0		c	$Al(OH)_4^-$
4	>8.5		d	$Al(OH)_3$

5.

	A (Reduction process)			B (Reduction potential (mV))
1	O_2 to H_2O		a	-120 to -180
2	NO_3 to N_2		b	-200 to -280
3	Fe^{3+} to Fe^{2+}		c	$+380$ to $+320$
4	SO_4^{2-} to S^{2-}		d	$+180$ to $+150$
5	CO_2 to CH_4		e	$+280$ to $+220$

TRUE OR FALSE

1. The organic matter contains the highest amount of nitrogen.

2. The chemical bond present in water is covalent bonding.

3. Term colloid is given by Crick in 1861.

4. C:N ratio of microorganisms is 10:1.

5. pH has unit of moles/ L.

6. The normality of 1M H_3PO_4 is 2N.

7. In soil organic colloids, the ion exchange is due to phenolic group only.

8. In general soil organic matter content in Indian soil is $< 0.5\%$.

9. Vanselow's equation is used for homovalent exchange reactions.

10. In polyphenol theory of humic substances formation, polyphenols are converted into quinines by phenoloxidases.

11. In soil, Gapon's equation is widely used to describe CEC of soils.

12. Ion exchange phenomenon discovered by Adams and Holmes.

13. The surface area of the adsorbing surface can be calculated using Kerr equation.

14. The anion exchange capacity of the soil increase with decrease in pH.

15. The cation exchange capacity of cereals is highest among all crops.

16. The stern layer is a diffused layer in the diffuse double layer concept.

17. Humic acid precipitates at 5.0 pH.

18. Electrophoresis is the migration of colloidal particles under the influence of electric field.

19. The specific conductivity of 0.01 N KCl solution at 25^{0-} C is 1.41dS/m.

20. The calomel electrode in measuring pH contains Hg_2Cl_2.

KEY TO CHOOSE THE CORRECT ANSWER

1. c	2. a	3. b	4. a	5. a	6. c
7. c	8. a	9. b	10. a	11. a	12. b
13. a	14. d	15. b	16. a	17. b	18. a
19. a	20. a	21. a	22. a	23. d	24. d
25. b	26. c	27. a	28. c	29. a	30. d
31. a	32. b	33. a	34. a	35. b	36. c
37. d	38. b	39. b	40. d	41. c	42. a
43. a	44. d	45. b	46. b	47. d	48. d
49. b	50. b	51. b	52. a	53. a	54. a
55. a	56. b	57. c	58. b	59. b	60. d
61. b	62. b	63. b	64. d	65. d	66. a
67. c	68. d	69. b	70. d	71. b	72. d
73. d	74. b	75. a	76. d	77. b	78. c
79. a	80. b	81. b	82. b	83. b	84. a
85. d	86. a	87. c	88. a	89. a	90. b
91. b	92. d	93. c	94. c	95. a	96. b
97. b	98. d	99. a	100. b	101. b	102. b
103. d					

KEY TO FILL IN THE BLANKS

1. $cmol\ (e^-)/kg$
2. Negative
3. 6.0
4. Micelle
5. 5.5
6. 13
7. 6
8. Increase of CEC
9. Minimum free energy
10. NH_4
11. Zeta potential
12. High swelling
13. Humin
14. 700000
15. Normality
16. 140:10:1.3:1.3
17. Space lattice
18. C.E Marshall
19. Decrease
20. Poise

KEY TO MATCH THE FOLLOWING

1. 3-a   5-b   1-c   2-d   e-4
2. 3-a   1-b   2-c
3. 3-a   4-b   1-c   2-d
4. 2-a   1-b   4-c   2-d
5. 4-a   5-b   1-c   3-d   2-e

KEY TO TRUE OR FALSE

1. False [True: carbon]
2. True
3. False [True: Thomas Graham]
4. False [True: 5:1]
5. False [True: Unitless]
6. False [True: 3N]
7. False [True: Phenolic and carboxylic]
8. True
9. False[True: Kerr's and Vanselow's equation]
10. True
11. True
12. False [True: Thomson and Way]
13. False [True: BET equation]
14. True

15. False [True: Legumes crops have higher due to more attraction of divalent by legumes]

16. False [True: it is fixed layer]

17. False [True: 4.8 pH]

18. True

19. True

20. True

<table><tr><td>6</td><td>SOIL MICROBIOLOGY</td></tr><tr><td></td><td>Usha Kumari and Prathibha, K</td></tr></table>

6 SOIL MICROBIOLOGY

Usha Kumari and Prathibha, K

CHOOSE THE CORRECT ANSWER

1. Dehydrogenase activity is a good index of biological activity of soil because it plays role in
    (a) Carbon metabolism
    (b) Respiration
    (c) Synthesis of call macromolecules
    (d) Cell division

2. Association of two microorganisms in which one is benefited while other remains unaffected is known as
    (a) Commensalism
    (b) Protoco operation
    (c) Ammensalism
    (d) Competition

3. Which group of microorganism is most active at the terminal stage of composting?
    (a) Actinomycetes
    (b) Bacteria
    (c) Fungi
    (d) Protozoa

4. The presence of 'Hartig net' is a characteristic of
    (a) Endomycorrhizae
    (b) Ectomycorrhizae
    (c) Arbuscular mycorrhizae
    (d) Ericoid mycorrhizae

5. Rhizosphere is determinate by
    (a) Gram positive spore forming bacteria
    (b) Gram negative spore forming bacteria
    (c) Gram negative non spore forming bacteria
    (d) Gram positive non spore forming bacteria

6. The toxicity of oxygen for strict anaerobe is due to the production of
    (a) Hydroxyl radical
    (b) Superoxide radical
    (c) Hydrogen peroxide
    (d) Cyanide

7. Toxicity of oxygen for strict anaerobe is due to the production of
    (a) Hydroxyl radical
    (b) Superoxide radical
    (c) Hydrogen peroxide
    (d) Cyanide

8. The specificity between legume host and Rhizobium spp. is governed by
   (a) Flavonoids
   (b) Tryptophan
   (c) Polysaccharides
   (d) Indole acetic acid

9. Some microorganism produce organic ligands to chelate iron in the soil and make it available to the plants. These organic ligands are known as
   (a) Porphyrins
   (b) Siderophores
   (c) Chromophores
   (d) Ferrodoxins

10. Aspergillus awamori is a
   (a) K-mobilizer
   (b) Organic matter decomposer
   (c) S-oxidizer
   (d) P- solubilizer

11. In India the work on biogas was first carried out by
   (a) J.N.Mukherjee
   (b) S.V.Desai
   (c) N.P.Datta
   (d) N.N.Goswami

12. 'cfu' is a unit for measuring
   (a) Soil pollution
   (b) Metal contamination
   (c) Soil quantity
   (d) Microbial population

13. Rhizobium japonicum fixes N in symbolic relationship with
   (a) Pea group
   (b) Lupin group
   (c) Soybean group
   (d) Phaseolus group

14. Prokaryotes in soil include the following groups of organisms
   (a) Fungi, Actinomycetes, bacteria, algae
   (b) Bacteria, archaea, actinomycetes, BGA
   (c) Protozoa, bacteria, fungi , archaea
   (d) Algae, protozoa, fungi, bacteria

15. Which one of the following organism is an obligate chemoautotroph?
   (a) Frankia
   (b) Nitrosomonas
   (c) Rhizobium
   (d) Anabaena

16. Chief constituents of gobar gas are
   (a) Methane and CO_2
   (b) Methane and CO
   (c) Ethylene and O
   (d) Ethylene and CO

17. CNG (Compressed Natural Gas) mainly constitutes of
   (a) Methane   (b) Ethylene   (c) Acetylene   (d) Oxygen

18. Gas filled in the fire extinguisher
   (a) Oxygen   (b) Carbon dioxide   (c) Nitrogen   (d) Hydrogen

19. The conversion of ammonium-N (NH_4^+) to nitrite-N (NO_2^-) by *Nitrosomonas*.
    (a) Ammonification
    (b) Hydrolysis
    (c) Nitrosation
    (d) Nitration

20. The conversion of nitrite (NO_2^-) to nitrate (NO_3^-) by *Nitrobacter*.
    (a) Ammonification
    (b) Aminization
    (c) Nitrosation
    (d) Nitration

21. The intermediate product of urea hydrolysis is
    (a) Carbonic acid
    (b) Ammonium carbonate
    (c) Ammonium carbonates
    (d) Ammonium carbamate

22. The essential component of urease enzyme is
    (a) Molybdenum
    (b) Nickel
    (c) Cobalt
    (d) Zinc

23. Conversion of protein to $R-NH_2$
    (a) Aminization
    (b) Ammonification
    (c) Nitrification
    (d) Nitration

24. ______ is the essential component of RNA Polymerase and required for the synthesis of m-RNA
    (a) N
    (b) Zn
    (c) Mn
    (d) Cu

25. Which of the following microorganism causes fatal poisoning in canned fruits and vegetables?
    (a) *Aspergillus flavus*
    (b) *Penicillium digitatum*
    (c) *Clostridium botulinum*
    (d) *Rhizoctonia solani*

26. Actinomycetes belong to
    (a) The fungi
    (b) Eukaryote
    (c) *Mycelia sterilia*
    (d) None of the above

27. The term 'Heterosis' was coined by
    (a) G.H. Shull
    (b) W. Bateson
    (c) T.H. Morgan
    (d) E.M. East

28. Organism most tolerant to soil moisture stress is
    (a) Fungi
    (b) Protozoa
    (c) Bacteria
    (d) Actinomycetes

29. In leguminous crops, *Rhizobium* enters root hair dissolving cell wall by
    (a) Cellulase
    (b) Chitinase
    (c) Pectinase
    (d) Lignolytic enzyme

30. Which of the following is P solubilizing bacteria?

    (a) *Pseudomonas striata*   (b) *Nitrosomonas*

    (c) *Thiobacillus ferroxidanse*   (d) *Fusarium oxysporum*

31. Methanogenesis in waterlogged soil takes place at Eh of

    (a) < -200 mV   (b) 200-300 mV

    (c) 300-400 mV   (d) > 400 mV

32. The microorganism predominant in flooded rice soils is

    (a) Bacteria   (b) Fungi

    (c) Actinomycetes   (d) Protozoa

33. Nitrogen reduction in symbiotic systems takes place at

    (a) Nodules surface   (b) Membrane envelop

    (c) Bacteroids   (d) Cytochrome

34. The 'Bangalore method' of composting is characterized by

    (a) Initial aerobic phase followed by anaerobic phase

    (b) Continuous aerobic phase

    (c) Initial anaerobic followed by aerobic phase

    (d) Continuous anaerobic phase

35. *Bradyrhizobium* sp. is suitable for

    (a) Groundnut   (b) Soybean   (c) Cowpea   (d) Alfalfa

36. Aerobic bacteria can assimilate only ___________ % of the carbon during
    organic matter decomposition

    (a) $< 1.0\%$ (b) $5 - 10\%$ (c) $20 - 40\%$ (d) $> 40\%$

37. The micro-organism most active at the terminal stage of composting are

    (a) Bacteria   (b) Fungi

    (c) Protozoa   (d) Actinomycetes

38. Which organism is leaf inhabitant?

    (a) Azotobacter   (b) Beijierankia   (c) Aerobacter (d) Rhizobium

39. The ratio of microbial biomass C to basal soil respiration is termed as

    (a) Microbial quotient   (b) Metabolic quotient

    (c) Mineralization index   (d) Labile C

40. Cunninghamella test is done to assess the organisms for its

    (a) N_2 – fixing capacity   (b) P – solubiliting capacity

    (c) S – oxidiziag capacity   (d) F – oxidixing capacity

41. What is the relationship between pH and Eh?

    (a) Eh = pH/0.0591
    (b) pH + Eh = 0.0591
    (c) pH = Eh/0.0591
    (d) Eh – 0.0591 = pH

42. Mineralization of soil nitrogen occurs when C:N ratio of organic matter is

    (a) < 20:1   (b) 20 – 30:1   (c) >30:1   (d) > 300:1

43. C:N ratio of well decomposed compost is

    (a) 100:1   (b) 20:1   (c) 50:1   (d) 10:1

44. Humic acid precipitates at the pH

    (a) 5.0   (b) 5.2   (c) 4.8   (d) 4.5

45. Maximum nitrogen to soil by

    (a) *Rhizobium meliloti*
    (b) *Rhizobium leguminosarum*
    (c) *Rhizobium japonicum*
    (d) *Rhizobium trifoli*

46. Which of the following is an stem nodulatingbacteria ?

    (a) *Sesbania rostrata*
    (b) *Rhizobium leguminosaram*
    (c) *Rhizobium japonicum*
    (d) *Rhizobium trifoli*

47. BGA as Nitrogen fixing was demonstrated in India by

    (a) P.K.Dey   (b) R.N.Singh   (c) Jenny   (d) Bousingault

48. Who introduced the Indore Method of compost preparation?

    (a) Hutchinson and Richards
    (b) Wiley
    (c) Howard
    (d) Richard

49. Epigeic earthworms

    (a) Make burrow in soil
    (b) Do not make burrow in soil
    (c) Are subsoil dwellers
    (d) Are found in soil

50. How much percentage of moisture is maintained in finished vermicomposting?

    (a) 10 – 20 %   (b) 30 – 40 %   (c) 50 – 60 %   (d) 80 – 90 %

51. Which of the following statement is not true regarding arbuscules of VAM fungi?

    (a) Transfer mineral nutrients from the fungi to host plant
    (b) Transfer sugar from host plant to the fungi
    (c) Work as storage organ for the fungi
    (d) A highly branched structure

52. Who for the first time showed that legumes can obtain nitrogen from air?

    (a) M. W. Beijerink
    (b) J. (B) Bousingault
    (c) Robert Koch
    (d) Louis Pasteur

53. What is the amount of N fixer by Azolla?
  (a) 20 – 25 kg/ha
  (b) 25 – 50 kg/ha
  (c) 25 – 30 kg/ha
  (d) 50 kg/ha

54. *Azotobacter* is a
  (a) Nitrogen fixing fungi
  (b) Nitrogen fixing bacteria
  (c) Phosphate solubilizing bacteria
  (d) Symbiotic bacteria forming Nodules on roots

55. What is the amount of nitrogen fixed by blue green algae?
  (a) 25 – 40 kg/ha
  (b) 25 – 30 kg/ha
  (c) 15 – 45 kg/ha
  (d) 15 – 30 kg/ha

56. Who were the first to try electron microscopy in the study of humic acids?
  (a) Tan and Giddens (1972)
  (b) Tan and MacCreery (1970)
  (c) Flaig and Beutelspachev (1951)
  (d) Schnitzer and Khan (1972)

57. According to Flaig hypothesis to what should be the source, or starting point, for the formation of fulvic and humic acids.
  (a) Lignin
  (b) Phenols
  (c) Quinone
  (d) Quinhydrone

58. Microorganism that has symbiotic association with casuarinas in atmospheric nitrogen fixation is
  (a) *Rhizobium*
  (b) *Alnus*
  (c) *Frankia*
  (d) *Azospirillum*

59. Organism (s) responsible for the conversion of Fe^{2+} to Fe^{3+} is/are
  (a) Ferrobacillus
  (b) Gallionella
  (c) Hydrogenomonas
  (d) Both a and b

60. Major constitute of natural gas
  (a) N_2
  (b) CH_4
  (c) CO_2
  (d) Argon

61. Which micro organism thrives well in acidic condition?
  (a) Bacteria
  (b) Fungi
  (c) Actinonycele
  (d) Algae

62. Soil has maximum population under normal condition
  (a) Thermopiles
  (b) Mesophiles
  (c) Psychrophiles
  (d) Macro fuses

63. For autotrophs source of carbon is
  (a) $AimCO_2$
  (b) Organic compound
  (c) Inorganic compounds
  (d) Sunlight

64. Acetobacter is associated with
    (a) rice   (b) maize   (c) sugarcane   (d) legumes

65. What is the percentage of organic carbon in organic matter?
    (a) 48   (b) 58   (c) 78   (d) 68

66. What is the C:N ration of Indian soils?
    (a) 30 - 32:1   (b) 100:1   (c) 20 – 22:1   (d) 10 – 12:1

67. The term humic acid was given by
    (a) Oden   (b) Miescher   (c) Berzelius   (d) Mulder

68. Fulvic acid is soluble in
    (a) Acid and Alkali   (b) Acid and Alcohol
    (c) Alkali and Alcohol   (d) Water

69. Individual fungal filaments are called
    (a) Hyphae   (b) Mycellium
    (c) Pseudopodia   (d) Flagella

70. Mycorrhizae is a symbiotic association between
    (a) Bacteria and plant bacteria
    (b) Fungi and bacteria
    (c) Two species of fungi
    (d) Fungi and roots of higher plants

71. Aflatoxin is a mycotoxin on produced by
    (a) *Bacillus polymyxa*   (b) *Aspergillus awamorii*
    (c) *Pseudomonas striata*   (d) *Apergillus flavus*

72. Bacteria from root nodules of legume was isolated for the first time by
    (a) J.(B) Bousingault   (b) M.W. Beijerink
    (c) Robert Koch   (d) Louis Pasteur

73. Penicillin was discovered by
    (a) Robert Koch   (b) Alexander Fleming
    (c) Waksman Fleming   (d) Winogradsky

74. The precursor of Indole Acetic Acid is
    (a) Allophane   (b) Tryptophane
    (c) Ethylene   (d) None of the above

75. Which of the following plays a role in ripening of fruits?
    (a) Auxin   (b) Cytokinin
    (c) Ethylene   (d) None of the above

76. The anaerobic breakdown of protein rich material is termed
    (a) Putrefaction
    (b) Denitrificcation
    (c) Depolymerization
    (d) None of these

77. Denitrification may also be termed as
    (a) Nitrate assimilation
    (b) Nitrate respiration
    (c) Denitrification
    (d) All of the above

78. The most common contaminant of rhizobium isolated from legume root legume root nodule is
    (a) *Agrobacterium*
    (b) *Azotobacter*
    (c) *Azospirillum*
    (d) None of the above

79. BGA was discovered as a nitrogen fixer in paddy field by
    (a) P.K.Dey
    (b) Beijerink
    (c) Fred
    (d) None of the above

80. *Rhizobium meliloti* causes nodulation in
    (a) Sweat clover (melilotis)
    (b) Alfalfa (Medicago)
    (c) Fenugreek (Trigonella)
    (d) All the above

81. Who is the 'Father of field experiment'?
    (a) M.W.Beijerink
    (b) J.B.Bousingault
    (c) Theoder de Saucer
    (d) Arnon and Stout

82. What is the temperature range of mesophiles?
    (a) $< 10°C$
    (b) $10 - 20°C$
    (c) $20 - 40°C$
    (d) $> 40°C$

83. What is the optimum pH for nitrification?
    (a) $5.5 - 6.5$
    (b) $6.5 - 7.5$
    (c) $7.5 - 8.5$
    (d) $5.0 - 7.0$

84. Which of the following is an aquatic fern?
    (a) *Azolla*
    (b) Anabena
    (c) BGA
    (d) Both a and b

85. Earthworm casts are rich in nutrients and   .
    (a) Bacteria
    (b) Fungi
    (c) Viruses
    (d) All

86. Soil viruses are
    (a) Obligate parasites
    (b) Facultative parasites
    (c) Saprophytes
    (d) Both a and b

87. Which microorganism is responsible for nitrification or nitrate formation?
    (a) Nitrosomonas
    (b) Nitrobacter
    (c) Clostridium
    (d) Nitrosococcus

88. Cellulose is a ___________
    (a) Monosaccharide
    (b) Disaccharide
    (c) Polysaccharide
    (d) None

89. The process of conversion of organic nitrogen into ammonia is known as ____________

    (a) Aminisation
    (b) Ammonification
    (c) nitrification
    (d) minerlization

90. The conversion factor organic carbon into organic matter is ___________
    (a) 1.65   (b) 2.78   (c) 1.724   (d) 1.427

91. Sucrose yields on hydrolysis ____________
    (a) Glucose and fructose
    (b) Galactose
    (c) Ribose
    (d) None

92. Fat is soluble in solvents
    (a) Polar   (b) Non – polar   (c) Unipolar   (d) None

93. Nitrogen biofertilizer for wheat is
    (a) *Azotobacter*
    (b) *Pseudomonas*
    (c) *Rhizobium*
    (d) None of these

94. Cyanobacteria that lacks heterocyst,can fix, nitrogen only under
    (a) Aerobic condition
    (b) Anaerobic condition
    (c) Normal condition
    (d) Dry condition

95. The fraction of organic matter is comparatively resistant to microbial degradation is
    (a) Hemicelluloses (b) Protein   (c) Cellulose   (d) Lignin

96. The normality of a solution of sodium hydroxide 100 ml of which contains 4g sodium hydroxide is
    (a) 0.1   (b) 1.0   (c) 4.0   (d) 0.4

97. Lignin theory was proposed by
    (a) Waksman   (b) Stevenson   (c) Kononova (d) Jenny

98. What is the range of pH in pH sacle?
    (a) 0 – 14
    (b) 1 – 14
    (c) – 14 to + 14
    (d) None

99. Which of the following is used as microbial culture for compost making?
    (a) Cellulolytic fungal culture
    (b) Phosphotika
    (c) Azotobacter culture
    (d) All

100. Possible indicator (s) of changes in soil quality is/are

    (a) Soil fertility and nutrient balance

    (b) Soil physical conditions

    (c) Soil organic Matter status

    (d) All the above

101. A standard solution is one whose strength is

    (a) Known   (b) Unknown

    (c) Both a and b   (d) None

102. What is the specific conductivity of KCl solution at 25°C?

    (a) 1.4118 dS/m   (b) 14.118 dS/m

    (c) 0.14118 dS/m   (d) None

103. Buffer solution for pH measurements include

    (a) pH 4.0   (b) pH 7.0   (c) pH 9.2   (d) All

104. In the leguminous plants, the nitrogen fixing enzymes are located in

    (a) Nodules of root   (b) Bacteria

    (c) Plant surface   (d) All

105. Nitrogen fixing filamentous bacteria is

    (a) *Rhizobium sp*   (b) *Azotoacter*

    (c) *Frankia*   (d) *Pseudomonas*

106. Agar agar is extracted from

    (a) Red algae   (b) Green aglae

    (c) Blue green algae   (d) None

107. Aerobic nitrogen fixing bacteria is

    (a) *Azospirillum*   (b) *Methanococcus*

    (c) *Clostridium*   (d) *Azotobacter*

108. Methane content of biogas varies from

    (a) 50 – 60%   (b) 40 – 45 %   (c) 80 – 85%   (d) 60 – 70%

109. An positive interaction in which one organism either gets benefit nor harmed this kind of interaction is known as

    (a) Neutralism   (b) Amensalism

    (c) Symbiosis   (d) Commensalism

110. The pH of the solution is 12, then the [OH⁻] ion concentration will be

    (a) 10^{12} (b) 0.02   (c) 0.0002   (d) 10^{-12}

111. Oxygen is toxic to nitrogenase because
    (a) Competitor for e-
    (b) Denatures disulphide bond
    (c) Reduces n fixation
    (d) Both a and b

112. The following organism used in mine extraction
    (a) *Thiobacillusthiooxidans*
    (b) *Ferro Bacillus sp*
    (c) *Bacillus sp*
    (d) *Agaricussp*

113. The source of oxygen for denitrifying bacteria for metabolism is
    (a) Atmosphere
    (b) Water
    (c) CO_2
    (d) Reduction of nitrates

114. Which of the following elements is essential for nodulation?
    (a) Mn
    (b) Ca
    (c) Co
    (d) B

115. In leghaemoglobin, the globulin part is produced by
    (a) Bacteriods
    (b) Host plants
    (c) Both a and b
    (d) None

116. Which of the following is the indicator of air pollution?
    (a) E.coli
    (b) Lichens
    (c) Fungi
    (d) Algae

FILL IN THE BLANKS

1. _____________ % lignin content in humus.

2. Term 'Rhizosphere' was coined by ___________.

3. Cellulose is______________.

4. Redox potential for emission of methane in waterlogged soils ___________.

5. Symbiotic N fixation (legume-Rhizobium) accounts for ________.

6. ______________ microorganism is not harmful to plants.

7. Most of the vaccines used now a days are produced from ___________.

8. Microorganism used for as a biological tool for evaluation of soil fertility of P, Mg, K ______________.

9. VAM (Vesicular Arbuscular Mycorrhiza) is ____________.

10. If CEC (Cation Exchange Capacity) increases the buffering capacity of soils ___________ first and then decreaseswith time.

11. Rhizobium enters root cells by dissolving _______________.

12. High C:N ratio results into _______________.

13. Heterocyst is associated with ____________.

14. Symbiotic association of roots of higher plants with fungus called ____________ helps in exploring soil by root interception for nutrition.

15. Arrange the following in the increasing order of their abundance ____________________.

16. Symbiotic association of fungus with algae is __________.

17. Which of the following bacteria oxidizes iron (conversion of ferrous to ferric) __________________.

18. Compute base saturation, if CEC is 20 cmol $(p^+)kg^{-1}$ and $[H^+]$ is 5 cmol $(p^+)kg^{-1}$____

19. Who discovered the symbiotic association between legumes and bacteria ______________.

20. An example of associative nitrogen fixer ____________.

21. Name the irregular structure that a bacteria forms inside the nodule ________________.

22. Which is the microorganism involved in the denitrification ____________.

23. Maximum denitrification occurs in ____________.

24. Conversion of N_2 to NO_3 is ____________.

25. Host specificity of Rhizobium to legume is due to ____________.

26. ________________ country most of the N required for crop production is obtained through N_2 fixation by legumes.

27. For autotrophic source of carbon is __________________.

28. Ooze test is done to detect ______________.

29. Biogas production is high from material rich in ______________.

MATCH THE FOLLOWING

1.

	A			B
1	Photoautrophs	a		Energy from sunlight C from organic matter
2	Chemoautotrophs	b		Energy from oxidation of inorganicsubstances, C from CO_2
3	Anaerobes	c		Exit as free living organism but fix N_2
4	Non- symbiotic N_2 fixers	d		Energy from sunlight, C from CO_2
5	Photoheterotrophs	e		Can use chemical species other than freeoxygen as electron acceptors

KEY TO CHOOSE THE CORRECT ANSWER

1. a	2. a	3. a	4. b	5. b	6. b
7. b	8. a	9. b	10. d	11. b	12. d
13. c	14. b	15. b	16. a	17. a	18. b
19. c	20. d	21. b	22. b	23. a	24. b
25. c	26. d	27. a	28. d	29. d	30. a
31. a	32. c	33. c	34. a	35. b	36. b
37. d	38. b	39. b	40. b	41. c	42. a
43. d	44. c	45. a	46. a	47. b	48. c
49. a	50. c	51. c	52. b	53. c	54. b
55. c	56. c	57. a	58. c	59. d	60. b
61. b	62. b	63. a	64. c	65. b	66. d
67. c	68. a	69. a	70. d	71. d	72. b
73. b	74. b	75. c	76. a	77. b	78. a
79. a	80. d	81. b	82. c	83. b	84. a
85. a	86. a	87. b	88. c	89. b	90. c
91. a	92. b	93. a	94. b	95. d	96. b
97. a	98. a	99. d	100. d	101. a	102. a
103. d	104. d	105. a	106. a	107. d	108. a
109. d	110. b	111. d	112. b	113. d	114. c
115. c	116. b				

KEY TO FILL IN THE BLANKS

1. 10-30%	2. Hiltner
3. Polysaccharides	4. −200mV
5. 140kgN/ha fixation	6. *Rhizobium*
7. *Actinomycetes*	8. *Azotobacter*
9. P mobilizer	10. Increases
11. Chitin	12. Immobilization of N

13. *Cyanobacteria*   14. Mycorrhiza

15. Bacteria/Fungi/ Actinomycetes (B>F>A)

16. Lichens   17. *Thiobacillus*

18. 75%   19. Beijerinck

20. *Azospirillum*   21. Bacteriod

22. *Pseudomonas*   23. Waterlogged soils

24. Biological oxidation   25. Flavinoids

26. New Zealand   27. Atmospheric CO_2

28. Bacterial diseases   29. Cellulose and hemicelluloses

KEY TO MATCH THE FOLLOWING

1.   1-a   2-e   3-d   4-b   5-c

<table><tr><td>7</td><td>

ANALYTICAL CHEMISTRY AND INSTRUMENTATION

Adiveppa Mallappa Asangi and M. Raghavendra Reddy
</td></tr></table>

CHOOSE THE CORRECT ANSWER

1. Proton donors are ___________________
   (a) Bases   (b) Acids
   (c) Both a and b   (d) Salts

2. Acids turn litmus paper to __________
   (a) Blue   (b) Red
   (c) Green   (d) Any of the colour

3. Acids react with bases to form___________
   (a) Salt alone   (b) Salt + water
   (c) Water alone   (d) None of these

4. _____________ turns phenolphthalein to pink
   (a) Weak acids   (b) Strong acids   (c) Both   (d) Alkalies

5. KH_2PO_4 is a ___________________
   (a) Acidic salt   (b) Neutral salt   (c) Basic salt   (d) All of these

6. $Bi(OH)_3$ is a ___________
   (a) Acid salt   (b) Neutral salt   (c) Basic salt   (d) All of these

7. $K_2Cr_2O_7$ is a ___________ agent
   (a) Oxidising   (b) Reducing   (c) Both   (d) None of these

8. Among the following, which one is a reducing agent
   (a) H_2O_2 (b) $K_2Cr_2O_7$
   (c) Ferrous sulphate   (d) $KMnO_4$

9. ___________ is an example of primary standard chemical
   (a) HCl   (b) H_2SO_4 (c) Na_2CO_3 (d) NaOH

10. $K_2Cr_2O_7$ is a _______________ standard chemical
    (a) Primary   (b) Secondary   (c) Tertiary   (d) a & b

11. 40 g of NaOH in a liter of water gives ___________ solution
    (a) $1\,M$ (b) $2\,N$ (c) 1 %   (d) 1 ppm

12. One gram equivalent weight of a substance in 1000 ml solution is

    (a) $1\ N$ (b) $1\ M$ (c) $1\ F$ (d) $1\ \%$

13. Equivalent weight = __________

    (a) $\dfrac{M.W.T}{Valency}$

    (b) $\dfrac{Valency}{M.W.T}$

    (c) $\dfrac{M.W.T \times 100}{Valency}$

    (d) $\dfrac{Valency \times 100}{M.W.T}$

14. __________ method is based on mass of substance

    (a) Volumetric   (b) Gravimetric

    (c) Specrometric   (d) Polarometric

15. __________ analysis is based on temperature

    (a) Potentiometry   (b) Conductometry

    (c) Thermal conductivity   (d) Calorimetry

16. Scattering of radiation can be applied in __________

    (a) Flamephotometry   (b) Polarometry

    (c) Turbidometry   (d) Reflectometry

17. Mass to charge ratio is applied in __________

    (a) Mass spectrometry   (b) XRD

    (c) AAS   (d) NMR

18. Fat or oil extraction is mainly done by __________ process

    (a) Solvent extraction   (b) Precipitation

    (c) Distillation   (d) Electrophorisi

19. Electrophorosis is mainly applied to __________ molecular weight compounds

    (a) High   (b) Low

    (c) Both a & b   (d) Can't be applied

20. Crystallites present along with colloids are mainly separated by __________

    (a) Electrophorosis   (b) Distillation

    (c) Dialysis   (d) Chromatography

21. __________ is indicator used for measuring dissolved oxygen

    (a) Ferroin   (b) $K_2Cr_2O_7$ (c) Starch   (d) All of these

22. Ca analysis is a __________ method
    (a) Precipitation
    (b) Complexometry
    (c) Redox
    (d) Volumetric

23. Which of the following is a internal indicator __________
    (a) Methyl red
    (b) Potassium ferricyanide
    (c) Ferroin
    (d) $K_2Cr_2O_7$

24. Wet oxidation of OC estimation developed by __________
    (a) Walkley & Black
    (b) Jackson
    (c) Parker
    (d) Tandon

25. Size of sieve used to prepare soil sample for OC estimation
    (a) 2.0 mm
    (b) 0.2 mm
    (c) 1.0 mm
    (d) 0.5 mm

26. Glucose solution is used during estimation of __________
    (a) N
    (b) P
    (c) OC
    (d) K

27. Digestion mixture components are __________
    (a) K_2SO_4
    (b) $CuSO_4$
    (c) Se
    (d) All of these

28. Ratio of $K_2SO_4 : CuSO_4 :$ Se in digestion mixture is __________
    (a) 20:100:1
    (b) 100:1:20
    (c) 100:20:1
    (d) 1:20:100

29. In digestion mixture $CuSO_4$ & Se act as __________
    (a) Catalysts
    (b) Redox agent
    (c) Complexing agent
    (d) No action

30. Ammonia liberated during distillation is trapped in __________
    (a) Boric acid
    (b) Perchloric acid
    (c) Sulphuric acid
    (d) Hydrochloric acid

31. Devardas alloy is used to convert __________
    (a) NO_3-N to NH_4-N
    (b) NH_4-N to NO_3-N
    (c) NO_2-N to NO_3-N
    (d) NH_4-N to NO_2-N

32. End colour of Boric acid solution after distillation is __________
    (a) Maroon
    (b) Green
    (c) Wine red
    (d) No colour

33. Which of the following is a primary orthophosphate __________
    (a) HPO_4^{-2}
    (b) $H_2PO_4^-$
    (c) H_3PO_4
    (d) PO_4^{3-}

34. Bray 1 method of P estimation is used for __________ soils
    (a) Acidic
    (b) Calcareous
    (c) Neutral
    (d) All soils

35. Bray 1 extractant comprised of __________
    (a) $NH_4F + H_2SO_4$
    (b) $HCl + H_2SO_4$
    (c) $NH_4F + HCl$
    (d) $H_3PO_4 + NH_4F$

36. Blue colour development in P estimation is done by _____________
    (a) Oxalic acid   (b) Malic acid   (c) Citric acid   (d) Ascorbic acid

37. pH of Olsen extractant is _______
    (a) 9.2   (b) 8.5   (c) 4.6   (d) 7.0

38. Activated charcoal must be used during estimation of _________
    (a) N   (b) P   (c) pH   (d) EC

39. Available form of potassium can be extracted through ____________
    (a) NH_4Cl (b) NH_4OAC (c) $(NH_4)_2 SO_4$ (d) NH_4OH

40. K can be estimated through
    (a) Turbidometer   (b) Spectrophotometer
    (c) Flame photometer   (d) All of these

41. Normality of NH_4OAC for estimation of K
    (a) $1\ N$ (b) $0.5\ N$ (c) $2\ N$ (d) $1.5\ N$

42. Buffer solutions used in standardisation of pH meter
    (a) 4   (b) 7   (c) 9.2   (d) All of these

43. Electrical conductivity of $0.01\ N$ KCl is __________ dS m^{-1}
    (a) 1.41   (b) 1.98   (c) 2.41   (d) 2.98

44. Orthophosphoric acid used in OC estimation for _______________
    (a) Sharp end point   (b) Buffer
    (c) Indicator   (d) Colouring agent

45. __________ % NaOH used for estimation of available N
    (a) 2.5   (b) 10   (c) 5   (d) 1

46. Flamephotometer works on _____________ of radiation by ions
    (a) Absorption   (b) Scattering   (c) Reflection   (d) Emission

47. Total quantity of -ve charges per unit weight of the soil is its _____________
    (a) AEC   (b) CEC   (c) Both a & b (d) None of these

48. Erichrome Black-T is used in estimation of
    (a) Ca alone   (b) Mg alone   (c) Ca + Mg   (d) S

49. _______ % $CaCl_2$ used in estimation of available S
    (a) 0.15   (b) 1.5   (c) 15   (d) 2.5

50. Extractant used in estimation of available B is ___________
    (a) 0.15 % H_2O_2 (b) 0.1 % H_2O_2 (c) Hot water   (d) Cold water

51. Shaking time for estimation of DTPA extractable micronutrients
____

    (a) 1 hr   (b) 2 hr   (c) 30 min   (d) 15 min

52. Hallow cathode lamp is used in ______________

    (a) AAS   (b) ICP-OES   (c) ICP-MS   (d) HPLC

53. Bartons reagent is used in analysis of plant ______________

    (a) N   (b) P   (c) K   (d) Ca

54. Plant samples should be dried at ________ ^0c

    (a) <60   (b) <80   (c) >60   (d) >80

55. Reagent used to estimate standards during 'P' estimation

    (a) KH_2PO_4 (b) K_2HPO_4 (c) PO_4 (d) H_2PO_3

56. Perchloric acid not advised to 'K' estimation if plant contains __________ % K

    (a) >1   (b) <1   (c) < 0.75   (d) < 0.5

57. N of AR grade H_2SO_4 is ______________

    (a) 30   (b) 36   (c) 17.4   (d) 14.3

58. Colour of phenolphthalein in acid range is ______________

    (a) Pink   (b) Blue   (c) Colourless (d) Pale Blue

59. Aqueous NH_3 contains ________ % of NH_3 by weight

    (a) 80   (b) 150   (c) 100   (d) 10

60. Buffer mask solution is used during estimation of __________

    (a) Cl   (b) Mo   (c) Si   (d) B

61. Crucial step in plant nutrient analysis is __________

    (a) Sampling   (b) Washing   (c) Drying   (d) Storing

62. Plant part to be sampled in wheat is ______________ for plant nutrient analysis

    (a) Flag leaf   (b) Ear leaf

    (c) Recently matured leaf   (d) All of these

63. Ideal time for plant sampling in sunflower is __________

    (a) Vegetative stage   (b) Flowering stage

    (c) Grain filling stage   (d) At any time

64. Triacid can't be used for estimation of ______

    (a) S   (b) K   (c) Ca   (d) P

65. Plant P is estimated at________nm of wavelength by Spectrophotometer
    (a) 340   (b) 420   (c) 520   (d) 660

66. Range of pH scale is ____________
    (a) 1-14   (b) 2-10   (c) 0-10   (d) 0-14

67. $1M$ HCl has pH of ____
    (a) 14   (b) 2   (c) 0   (d) 10

68. $1M$ NaOH has pH of ____
    (a) 14   (b) 10   (c) 12   (d) 1

69. Molecular weight of water is ____________
    (a) 16.016   (b) 18.016   (c) 15.016   (d) 10.016

70. Weight of one litre of water at 25 ^{0}C is _________
    (a) 1000 g   (b) 997 g   (c) 1003 g   (d) 990 g

71. Molar concentration of water is __________
    (a) 52.39   (b) 54.39   (c) 55.39   (d) 53.39

72. pH meter is developed by ________________
    (a) Beckman   (b) Sorenson   (c) Silen   (d) Jackson

73. Wavelength of visible light is ________ nm
    (a) 400 - 700   (b) 200 - 800   (c) 1 - 400   (d) 800 - 1000

74. Fuel used in flame photometer
    (a) Propane   (b) Hydrogen   (c) Acetylene   (d) All of these

75. ppm is a _________
    (a) mg/kg   (b) µg/kg   (c) mg/g   (d) g/kg

76. 1000 ppm = __________ %
    (a) 1   (b) 0.1   (c) 10   (d) 100

77. Wavelength region of red colour is ________ nm
    (a) 380 - 470   (b) 490 - 500   (c) 560 - 580   (d) 650 - 750

78. Cuvette used in IR Spectrophotometer is ________
    (a) Quartz   (b) Glass   (c) KCl   (d) NaCl

79. Cuvette used in UV Spectrophotometer is ____________
    (a) Quartz   (b) Glass   (c) KCl   (d) NaCl

80. Interference in AAS is/ are ______________ interferences
    (a) Chemical   (b) Ionisation   (c) Matrix   (d) All of these

81. Bragg's law is applied in ___________________

    (a) XRD   (b) DTA   (c) AAS   (d) All of these

82. "Each molecule has its own tendency to volatalise" is given by ___________

    (a) Tandon   (b) Jackson   (c) Henry   (d) Viets

83. Time taken by eluent from injector port to detector port is _______________

    (a) Retention time   (b) Dormant time

    (c) Holding time   (d) Flow rate

84. Principle in paper chromatography is ___________

    (a) Adsorption   (b) Partition

    (c) Ion exchange   (d) Exclusion

85. HPLC will be worked with mobile phase of a ___________

    (a) Solid   (b) Liquid   (c) Gas   (d) All of these

86. Chromatography can be used in _____________ analysis

    (a) Qualitative   (b) Quantitative   (c) Both   (d) None of these

87. ___________ detector called as work horse in environmental analysis

    (a) Thermal conductivity   (b) Flame photometer

    (c) Flame ionization   (d) Mass selective

88. Delayed fluorescence is called as _____________

    (a) Phosphorescence   (b) Luminescence

    (c) Scattering   (d) None of these

89. Which of the following is a Flourescent compound

    (a) Petroleum oil   (b) Quinine sulphate

    (c) Fluorspar   (d) All of these

90. Dry ashing using muffle furnace can be used to estimate

    (a) Plant – Boron   (b) Plant –Mo

    (c) Plant-Cu   (d) Plant-Ca

FILL IN THE BLANKS

1. Acids are _________ in taste.

2. Acids react with carbonates to form ________________.

3. ___________ are proton acceptors.

4. ___________ contains OH groups.

5. _____________ chemicals will give exact concentrations when known weight of sample dissolved in known volume of water.

6. Alkalies turn red litmus paper to _____________.

7. One gram in 100 ml is _________ % solution.

8. Known amount of solute in known amount of solvent is called as _________________.

9. One gram molecular weight of substance in 1000 g of solvent is a _____________.

10. _____________ analysis is a process of determining the amounts of each compound in a sample.

11. _____________ is process based on solubility differences between analyte and unwanted component.

12. _____________ is a common method of separation of liquid components from a mixture.

13. Equivalent weight of substance divided by 1000 gives _____________.

14. $ml \times N = $ _____________.

15. $(ppm / eq.wt) = $ _________________.

16. _____________ are the organic dyes or solutions which indicates end point of a chemical reaction.

17. _____________ is an indicator of Cl test.

18. EDTA is a _________________ agent.

19. $KMnO_4$ is a _____________ indicator.

20. _________________ is used as indicator during standardization of acid.

21. The distance between two successive maxima on the electro magnetic wave is called _________________.

22. _____________ & _____________ are components of mixed indicator used in N estimation.

23. Devardas alloy comprised of _________, _______, _______.

24. Bray-1 extractant contains _______ N NH_4F and _______ N HCl.

25. Olsen extractant contains ______ M $NaHCO_3$.

26. _____________ buffer solution used in estimation of Ca + Mg by versanate method.

27. The pH of the buffer for Ca is _______ and for Ca + Mg is _____ by complexometry.

28. Spiral augar is used for soil sampling in __________ soils.

29. Glass electrode coated with _____________.

30. H_2SO_4 enables the easy digestion of organic matter in the estimation of OC by rendering ____________________.

31. Glass beads used during distillation to estimate available N is to avoid ___________.

32. Beer - Lamberts law applied in ____________ instrument.

33. _______________ is used to make colloid very stable during estimation of available 'S'.

34. Azomethane - H used in estimation of ____________.

35. Available - S in soil measured using Spectrophtometer at _______ nm of wavelength.

36. During estimation of available - S, role of conditioning agent is ___________________.

37. Content of DTPA extractant are ___________, ___________ & ___________.

38. 1 nm = _______ Å.

39. 1 mm = _______ nm.

40. ____________ rays emitted due to movement of electrons close to nuclei of heavy atoms.

41. Wavelength of mocrowave region is ___________. (0.1 mm to 1 cm)

42. If 2 mg of a solute dissolved in a litre of water, its concentration is ______ ppm.

43. Light source for visible region spectrophotometry is _______________.

44. The light source for UV Spectrophotometer is ____________.

45. Zeeman effect can be seen in ____________ instrument.

46. Deuterium lamp is used in AAS for ____________________________.

47. ____________ is an instrument employed for chromatography.

48. __________ is fluid that enters into column of chromatograph.

49. Mobile phase in gas chromatography will be a __________.

50. Expand HPLC ____________.

MATCH THE FOLLOWING

1.

	A			B
1	Bray I - P	a	$0.5\ N$ HCl $+\ 0.025\ M$ H$_2$SO$_4$
2	AB DTPA - P	b	$0.03\ N$ NH$_4$F $+\ 0.025\ M$ HCl
3	Olsen - P	c	$0.73\ M$ NaO AC $+\ 7.4\ M$ HO AC
4	Mehlich- P	d	$0.5\ M$ NaHCO$_3$
5	Morgam - P	e	$2\ M$ NH$_4$HCO$_3$ $+\ 0.005\ M$ DTPA

2.

A (Element)			B (Extractant)	
1	P	a	DTPA
2	K	b	Hot water
3	S	c	Olsen
4	Fe	d	CaCl$_2$
5	B	e	$1N$ NH$_4$OAC

3.

A (Chemical)			B (Role in textural analysis)	
1	H$_2$O$_2$	a	Deffloculation
2	HCl	b	CaCO$_3$ removal
3	Calgon reagent (Sodium Hexa Metaphosphate)	c	Organic matter decomposition

4.

A (Instrument)			B (Scientist)	
1	pH meter	a	Alan Walsh
2	AAS	b	Beckman
3	Flame photometer	c	Tswett
4	Chromatography	d	Bunsen and Kirchhoff

5.

A (Instrument)			B (Element)	
1	AAS	a	P
2	Spectrophotometer	b	C
3	Flame photometer	c	Zn
4	Turbidometer	d	K
5	CHNS Analyzer	e	S

6.

	A (Principle)			B (Instrument)
1	Absorption		a	Flame photometer
2	Emission		b	Refractrometer
3	Scattering		c	Polarimeter
4	Rotation		d	Spectrophotometer
5	Diffraction		e	XRD
6	Refraction		f	Turbidometer

7.

	A (Parameter)			B (Unit)
1	Density		a	$dS\ m^{-1}$
2	pH		b	g/cc
3	EC		c	No units
4	Soil-N		d	Kg/ha
5	CEC		e	$m.eq\ L^{-1}$

8.

	A (Extractant)			B (pH)
1	DTPA		a	8.5
2	AB DTPA		b	3.3
3	Olsen reagent		c	7.3
4	Morgan reagent		d	4.8
5	Tamn reagent		e	7.6

TRUE OR FALSE

1. Alkalies are soluble in water.

2. One gram molecular weight of substance dissolved in one litre solution gives **one normal solution**.

3. Qualitative analysis is a process of identifying different compounds present in a sample.

4. Qualitative analysis gives **numerical information** of solute.

5. Spectrophotometer can be used for both qualitative and quantitative analysis.

6. Flame photometer can be used for **qualitative** analysis.

7. Radioactivity of an ion can be measured through its **interaction with radiation**.

8. Polarimetry measures optical rotation of substance.

9. Organic carbon estimation is an example of gravimetric method.

10. Chromatography is based on process of differential migration of a mobile phase relative to an immobile phase.

11. (m. eq x eq. wt) = mg

12. Starch is used as indicator to measure COD.

13. K_2CrO_4 is a precipitation indicator.

14. Phenolphthalein is used as indicator during standardization of boron.

15. Ferroin indicator is used in **N** estimation.

16. K_2SO_4 in digestion mixture will increase the boiling point of H_2SO_4.

17. Liquid paraffin added in distillation process is to avoid frothing.

18. Olsen method of P estimation used for **only neutral** soils.

19. **Equivalent points** represents the point of completion of reaction.

20. Example for complexometry is an estimation of **Ca alone**.

21. Hot water soluble B method was given by Berger and Truag.

22. TEA is used for buffering because it burns clearly during atomization in estimation of micronutrients.

23. Plant samples should be brought to lab in paper bags.

24. Triacid mixture contains $HNO_3:H_2SO_4:HClO_4$.

25. DTPA is used in estimation of **plant** micronutrients.

26. K emits an lilac colour when excited in Flamephotometer.

27. Murexide indicator is a mixture of ammonium purporate and K_2SO_4.

28. Normality of AR grade HNO_3 is 16.

29. In the titration between weak acid and weak base, the **suitable indicator is thymol blue.**

30. Sodium acetate - Acetic acid buffer solution is also called as Morgan's reagent.

31. AAS was developed by Alan Walsh.

32. H electrode is also called as indicator electrode.

33. Calomel electrode is also called as reference electrode.

34. 1 ml in a litre gives **one ppm**.

35. Monochromators are used for getting a single colour light.

36. Transmittance can be measured through $A = 2 - \log T$

37. Calorimeter cannot detect wider range light like UV and IR.

38. Wavelength of X rays is 10^{-10} meters.

39. Use of flame in Flame photometer is for both atomization and excitation.

40. Complementary colour for absorbed red colour is blue green.

41. Chromatography separates two physical mixtures.

42. Higher the flow rate lesser will be the retention time in chromatography.

43. Electrolytic conductivity detector is called universal detector.

44. Fluorescence is the phenomena of emission of longer wavelength than incident radiation.

45. Ca & Mg can be estimated by Versanate method.

46. Boron can be estimated by **AAS**.

47. OC can be estimated by both dry and wet combustion methods.

48. Plants and manure samples can be analyzed in similar ways for nutrient content.

49. Soil samples should not be collected near bunds of the fields.

50. Extractant used for Mo estimation is grigg's reagent.

KEY TO CHOOSE THE CORRECT ANSWER

1. b	2. b	3. b	4. d	5. a	6. c
7. a	8. c	9. c	10. a	11. a	12. a
13. a	14. b	15. c	16. c	17. a	18. a
19. a	20. c	21. c	22. b	23. a	24. a
25. b	26. c	27. d	28. c	29. a	30. a
31. a	32. b	33. b	34. a	35. c	36. d
37. b	38. b	39. b	40. c	41. a	42. d
43. a	44. a	45. a	46. d	47. b	48. c
49. a	50. c	51. b	52. a	53. b	54. a
55. a	56. a	57. b	58. c	59. c	60. d
61. a	62. a	63. b	64. a	65. b	66. d
67. c	68. a	69. b	70. b	71. c	72. a
73. a	74. a	75. a	76. b	77. d	78. d
79. a	80. d	81. a	82. c	83. a	84. b
85. b	86. c	87. c	88. a	89. d	90. a

KEY TO FILL IN THE BLANKS

1. Sour
2. $CO_2 + H_2O$
3. Bases
4. Alkalies
5. Primary standard
6. Blue

7. 1 % solution
8. Solution
9. One molal solution
10. Quantitative
11. Precipitation
12. Distillation
13. m.eq.wt
14. m.eq
15. m.eq/ lt
16. Indicators
17. K_2CrO_4
18. Chelating
19. Self
20. Methyl orange
21. Wave length
22. Bromocresol green & Methyl red
23. Cu, Al, Zn
24. 0.03 & 0.025
25. 0.5
26. $NH_4Cl + NH_4OH$
27. 12 & 10
28. Very hard
29. AgCl
30. Heat of dilution
31. Bumping
32. Spectrophotometer
33. Gum acacia
34. Available B
35. 420
36. To get uniform turbidity
37. 0.005 M DTPA, 0.01 M $CaCl_2$, 0.1 M TEA
38. 10
39. 10^6
40. X - rays
41. 0.1 mm to 1 cm
42. 2 ppm
43. Tungsten lamp
44. Deuterium lamp
45. AAS
46. To correct background emissions
47. Chromatograph
48. Eluent
49. Gas
50. High Performance Liquid Chromatography

KEY TO MATCH THE FOLLOWING

1.	4-a	1-b	5-c	3-d	2-e	
2.	4-a	5-b	1-c	3-d	2-e	
3.	3-a	2-b	1-c			
4.	2-a	1-b	4-c	3-d		
5.	2-a	5-b	1-c	3-d	4-e	
6.	2-a	6-b	4-c	1-d	5-e	3-f
7.	3-a	1-b	2-c	4-d	5-e	
8.	3-a	5-b	1-c	4-d	2-e	

KEY TO TRUE OR FALSE

1. True
2. False [True: One molar solution]
3. True
4. False [True: It never gives numerical information]
5. True
6. False [True: Quantitative]
7. False [True: Nuclear properties]
8. True
9. True
10. True
11. True
12. False [True: Ferroin]
13. True
14. True
15. False [True: Organic carbon]
16. True
17. True
18. False [True: Neutral & Alkaline]
19. False [True: End point]
20. False [True: Ca & Mg]
21. True
22. True
23. True
24. True
25. False [True: Soil]
26. True
27. True
28. True
29. False [True: No suitable indicator]
30. True
31. True
32. True
33. True
34. False [True: 1000 ppm]
35. True
36. True
37. True
38. True
39. True
40. True
41. True
42. True
43. False [True: Thermal conductivity]
44. True
45. True
46. False [True: Spectrophotometer]
47. True
48. True
49. True
50. True

8

PROBLEMATIC SOILS

P.N. Siva Prasad and C.T. Subbarayappa

FILL IN THE BLANKS

1. Soil pH measures _________ which is expressed in logarithmic terms

2. Activity is the product of ____________ × Activity coefficient

3. The concept of pH was given by __________

4. The product of H^+ and OH^- concentrations is _____________ of water

5. The hydrogen ions in the soil solution contribute to ____________

6. Acidity developed due to adsorbed hydrogen (H^+) and aluminium (Al^{+3}) ions on soil colloidsis _____________

7. The optimum pH for availability of the most of the nutrients is ________

8. The Hydrolysis of $Al(OH)^{+3}$ ion occurs in the pH range of-_________

9. The pH of extremely acid soils is----- ___________

10. Acid soils are formed in this region of ____________ areas

11. Acid soils are reclaimed by _____________ amendment

12. For the pH 5.0 and above, the main source of hydrogen in soils is ______________

13. For the pH below 5.0, the main source of hydrogen in soils is ______________

14. Neutralizing index (N.I) of the liming material if fineness factor is 70 and CCE is 90 % is _____________

15. The H^+ ion concentration has a __________ change between each whole pH number

16. Due to the high soil acidity ______&_______ became highly solubulized and available in toxic amounts

17. __________ and _________ ions are also called as potential determining ions

18. Total acid soils occupy about_______________ *i.e* total one fourth of total geographical area of the country

19. The acid sulphate soils are unique to_______________

20. The criteria used to classify salt affected soils are_____ & _______

21. The soils are said to be saline if they contain an excess of soluble salts and sodic if they contain an excess of _______________

22. The saline soils are predominant at composition of _____________ of Na, Ca and Mg

23. The alkali soils are dominant at composition _____________________of sodium

24. Solon chacks refers to __________________

25. Solonetz refers to ________________

26. Saline - alkali soils have been formed as a result of both ___________

____

27. Total soluble salt content of saline soils is _____________

28. Soils having ESP value > 15 are considered as ___________

29. Leaching requirement is calculated using the formula LR = ____

30. Leaching requirement in terms of EC is calculated using LR = _____

31. Gypsum contains 29.2 % Ca and __________sulphur

32. The passage of 1 m leaching water per metre of soil depth under continuous ponded conditions normally removes approximately ___________ of soluble salts from the soil

33. The most harmful salt in saline and alkali soils is____________

34. The ratio of osmotic potential (bars) and EC (dS/m) is always__________

35. The salt tolerant mechanism in dhainch is due to _____________

36. The salt tolerant mechanism in sugar beet is due to _____________

37. The columnar structure is mostly found in ________________

39. Salt affected soils occur predominantly in ___________

40. Electrical conductivity of soil is a direct measure of _____________

41. The most important influence of submerging a soil in water is to

____

42. Soils that are saturated with water for a sufficiently long time in a year to give the soil to give distinctive gley horizons resulting from oxidation reduction-process is ____________

43. _______________ is a quantitative measure of tendency of a given system to oxidize or reduce substances

44. ____________ is negative log of electron activity, a measure of electron activity

45. _________in irrigation water is expected having Mg : Ca ratio more than one

46. Salinity classes are indicated by ________________

47. Which class of saline water is suitable for irrigation purpose _________

48. Which class is not suitable for irrigation purpose ____________

49. Sodium Classes are indicated by ____________

50. Sodium hazardous water is classified based on ___________

51. _______________ is mainly used to predict 'Na' hazard and it is a relation between Na, Ca & $CaCO_3$ in the irrigation water

52. If Salt index is _________ it indicates high quality of irrigation water and if it is __________ indicates harmful for irrigation

53. ________ of boron is essential for irrigation water and can be used safely

54. _______________ refers to breaking down of soil aggregates at near saturation into ultimate soil particles

55. _______________ process of clay decomposition and transformation under the influence of periodic reduction of iron oxides to ferrous ions

56. The determination of pE value of redox potential (Eh) divided by _______________ will give the pE value

57. In submerged soils nitrogen mineralization stops at ____________

58. The important green house gas released during submerged condition is

____

59. Submergence of the soil increases the availability of ____________

60. Submergence of soil decreases the availability of _______________

61. The irrigation water with EC of 2350 μ mhos cm^{-1} and SAR of 17 is categorized under the class _________

62. The irrigation water with EC of 2350 μ mhos cm^{-1} and SAR of 25 is categorized under the class _________

63. As per USSL diagram based on SAR and EC, the irrigation water can be classified upto a maximum of ____________

CHOOSE THE CORRECT ANSWER

1. Resistance to chance in pH of a solution is
   (a) Active acidity
   (b) % Base Saturation
   (c) Buffering Capacity
   (d) Dissociation constant

2. For the determination of lime requirement, which methods is used
   (a) Schoemaker *et al.*
   (b) Schonovar method
   (c) $BaCl_2$ – TEA method (Mehlich)
   (d) Both A and C

3. In percent base saturation, BS (%)=(S/T)×100, S denotes
   (a) Basic cations
   (b) Total cations
   (c) Total anion
   (d) Total ions

4. Microbes which are thrive well in acid soils is
   (a) Fungi
   (b) Bacteria
   (c) Actinomycetes
   (d) Algae

5. What is the range of pH in pH Scale
   (a) 0-14
   (b) 1-14
   (c) 1-7
   (d) 0-7

6. Buffering capacity is ————— is more in which type of minerals
   (a) 1:1
   (b) 2:1
   (c) Both
   (d) 2:2

7. The ion Activity coefficient is calculated by
   (a) Donan equation
   (b) Debye - Huckel equation
   (c) Lewis and Randall
   (d) Van't Hoff equation

8. One unit of pH change at 25 °C is equivalent to change in potential difference of
   (a) 0.059V
   (b) 0.59V
   (c) 5.91V
   (d) 59.1 V

9. If the soil pH increase from 4 to 6, the H^+ ion concentration changes by
   (a) 1/20
   (b) 2/10
   (c) 1/10
   (d) 1/100

10. If P^{OH} of solution is 8, then its pH is
    (a) 14
    (b) 10
    (c) 6
    (d) 8

11. The Saline/ salt affected soils are placed in the order
    (a) Aridisols
    (b) Andisols
    (c) Ultisols
    (d) Alfisols

12. The ochric epipedon is found in which great soil group
    (a) Salorthids
    (b) Natrargids
    (c) Nadurargids
    (d) Both B & C

13. The formula for estimation of total dissolved solids (mg/l)
    (a) EC (dS/m)×640
    (b) EC (dS/m)×10
    (c) EC (dS/m)× (-0.36)
    (d) None of the above

14. The sum of the soluble cations or anions (mmol/L) is calculated using the formula (salt concentration)
    (a) EC (dS/m)×640
    (b) EC (dS/m)×10
    (c) EC (dS/m)× (-0.36)
    (d) None of the above

15. Exchangeable sodium percentage (ESP) is determined using the formula

    (a) $$ESP = \frac{Exchangeable\ Na^+ \times 100}{CEC}$$

    (b) $$ESP = \frac{Exchangeable\ Na^+ \times CEC}{100}$$

    (c) $$ESP = \frac{CEC \times 100}{Exchangeable\ Na^+}$$

    (d) None of the above

16. EC is expressed in terms of
    (a) Millmhos/cm
    (b) Decisiemens/ metre
    (c) meq/100 g
    (d) Both A & B

17. Among the following which is best suits for saline soil
    (a) pH < 8.50
    ESP <15
    EC < 4 dS/m
    (b) pH < 8.50
    ESP > 15
    EC >4 dS/m
    (c) pH > 8.50
    ESP > 15
    EC < 4 dS/m
    (d) pH >8.50
    ESP < 15
    EC > 4 dS/m

18. Choose the correct order of increasing flocculation of the following ions
    (a) $Ca^{+2}> Mg^{+2}> Al^{+3}> K^+> Al^{+3}$
    (b) $Mg^{+2}> Ca^{+2}> K^+>Na^+>Al^{+3}$
    (c) $Al^{+3} >Ca^{+2}> Mg^{+2}> K^+>Na^+$
    (d) $Al^{+3}>K^+> Na^+>Mg^{+2}> Ca^{+2}$

19. Among the following which best suits for alkali soils
    (a) pH >8.50
    ESP > 15
    EC < 4 dS/m
    (b) pH >8.50
    ESP > 15
    EC> 4 dS/m
    (c) pH < 8.50
    ESP < 15
    EC > 4 dS/m
    (d) None of the above

20. Saline soils are also called as
    (a) White alkali soils
    (b) Black alkali soils
    (c) Both A and B
    (d) None of these

21. Alkali soils are also called as
    (a) White alkali soils
    (b) Black alkali soils
    (c) Both A and B
    (d) None of these

22. Central soil salinity research institute is located at
    (a) Delhi
    (b) Mumbai
    (c) Chennai
    (d) Karnal

23. Which among the crop is tolerant to salinity
    (a) Sunflower
    (b) Barley
    (c) Cowpea
    (d) Groundnut

24. Most salt tolerant grass is
    (a) Karnal grass
    (b) Rhodes grass
    (c) Para grass
    (d) Bermuda grass

25. Halomorphic soils are evolved by which process,
    (a) Calcification
    (b) Gleization
    (c) Salinization
    (d) None of those

26. Who developed the method of gypsum requirement
    (a) Shoemaker
    (b) Schoonover
    (c) Stanford & English
    (d) Olsen

27. Which among the soils have poor physical condition
    (a) Sodic
    (b) Saline
    (c) Acidic
    (d) None

28. The salt tolerant mechanism in barley is due to
    (a) Exclusion
    (b) Osmoregulation
    (c) Salt dilution
    (d) None

29. The most widely used source of calcium to reclaim sodic soils is
    (a) Gypsum
    (b) Lime
    (c) FeS_2
    (d) None

30. A process in which accumulation of soluble salts particularly Cl^- SO_4^{2-} of Ca^{2+}, Mg^{++} Na^+ and K^+ takes places is called
    (a) Gleization
    (b) Laterization
    (c) Salinization
    (d) Calcification

31. If the soils EC value is 6-8, then the crop most suited is
    (a) Barley
    (b) Gram
    (c) Mustard
    (d) All of these

32. Which of the following pairs is correctly matched
    (a) Alkali soil pH >8.5
    (b) Saline soil pH< 8.5
    (c) Both (A) & (B)
    (d) None of the above

33. Which of the following tree species is used for the reclamation of salt affected soils.
    (a) Casurina
    (b) Prosopis
    (c) Eucalyptus
    (d) All of those

34. Sodium uptake potential of sugar beet crop is grouped as

    (a) High   (b) Low   (c) Medium   (d) Very low

35. Which of the following type of soil has low permeability

    (a) Acidic   (b) Alkali   (c) Saline   (d) None of these

36. Gypsum equivalent in Iron sulphate ($FeSO_4\ 7\ H_2O$)

    (a) 1.62   (b) 1.78   (c) 1.42   (d) 1.24

    37. Gypsum equivalent in sulphuric acid (H_2SO_4) is

    (a) 1.78   (b) 0.18   (c) 0.57   (d) 0.75

38. The following crop most sensitive to crop salinity is,

    (a) Wheat   (b) Barley   (c) Sugarbeet   (d) Chick pea

39. The soluble source of calcium that is used to reclaim alkaline/ sodic soils is,

    (a) Gypsum   (b) Calcium chloride

    (c) Calcite   (d) Both (a) & (b)

40. The gypsum requirement for amelioration of alkali/sodic soils depends upon

    (a) Exchangeable Na content

    (b) Exchange efficiency to be replaced

    (c) Depth of soil to be reclaimed

    (d) All the above

41. The main electrochemical changes that influence the chemistry and fertility of submerged soils and growing of cropssuch as wetland rice include:

    (a) A decrease in Eh or reduction of the soil.

    (b) An increase in pH of acid soils and a decrease in pH ofalkali soils

    (c) An increase in specific conductance and ionic strengthof soil solution.

    (d) Sorption–desorption reactionsand the availability of major and micronutrients.

    (e) All the above

42. High concentration of sodium water is undesirable for irrigation became it causes

    (a) De-flocculation   (b) Sealing the pores of soil

    (c) Make the soil impermeable   (d) A,B,C

43. Following is the horizon sequence of submerged soils

    (a) Oxidized A horizon – Mottled Zone - Permanently reduced B horizon

    (b) Permanently reduced B horizon – A horizon – O horizon

    (c) A horizon – B horizon – C horizon

    (d) None of the above

44. _______ Eh destroys NO_3^-N but favours NH_4^+N accumulation and nitrogen fixation

    (a) Low   (b) High   (c) Increased   (d) Constant

45. The availability of zinc decreases due to submergence may be attributed to the formation of

    (a) Frankalinite   (b) Sphalerite   (c) Smithsonite (d) All the above

46. EBT is used for the determination of

    (a) Lime requirement   (b) Gypsum requirement

    (c) Organic Carbon   (d) Boron

47. The sequential redox reactions in submerged soils is as follows

    (a) $O_2 > NO_3 = Mn^{4+} > Fe^{+3} > SO_4^{2-} > CO_2$

    (b) $CO_2 > SO_4^{2-} > Mn^{4+} > Fe^{+3}$

    (c) $Mn^{4+} > Fe^{+3} > SO_4^{2-} > CO_2 > O_2 > NO_3$

    (d) None

TRUE OR FALSE

1. Liming is always limited to neutralize active acidity and a part of exchangeable acidity.

2. By increasing salt concentration in an alkaline soil, the pH value is lowered due to reduction in the hydrolysis of the exchangeable cations like sodium.

3. Gypsum can be a source of sulphur for surface soil, but it can amend the surface acidity.

4. For a soil with subsoil acidity, gypsum or phosphogypsum is useful.

5. All micronutrients (Fe, Mn, Cu, Zn, Co) except Mo are not available in acidic pH.

6. In acid pH actinomycetes and bacteria are well affected only fungi will survive.

MATCH THE FOLLOWING

1.

A		B	
1	% B.S >80 %	a	Highly fertile soils
2	% B.S 50-80 %	b	Medium fertile soils
3	% B.S <50 %	c	Low fertile soils

2.

	A			B
1	Calcite	a		$CaCO_3$
2	Dolomite	b		CaO
3	Quick lime	c		$CaCO_3 \, MgCO_3$

3.

	A			B
1	Law of minimum	a		Sorensen
2	Phosphate potential	b		Shoemaker *et al*
3	pH	c		Liebig
4	Lime requirement	d		Schonover
5	Gypsum requirement	e		Schofield

4.

	A (Water class)			B EC (mSm^{-1})
1	Low salinity	a		0-25
2	Medium salinity	b		75-225
3	High salinity	c		25-50
4	Very high salinity	d		225-500

5.

	A (Water class)			B (SAR)
1	S1 Low Na^+ hazard	a		10-18
2	S2 Medium Na^+ hazard	b		0-10
3	S3 High Na^+ hazard	c		18-20
4	S4 Very high Na^+ hazard	d		> 26

6.

	A (RSC value ($m \, el^{-1}$))			B
1	<1.25	a		Water can be used safely
2	1.25-2.5	b		Water can be used with certain management
3	>2.5	c		Unsuitable for irrigation purpose

7.

	A (Chloride concentration ($m \, el^{-1}$))			B (Water quality)
1	4	a		Excellent water
2	4-7	b		Slightly usable
3	7-12	c		Moderately good water
4	12-20	d		Not suitable for irrigation purpose

8.

	A (Amendments)		B (Soil conditions)
1	Gypsum	a	Saline and alkali soils (pH upto 9)
2	Sulphur and Iron pyrites	b	Alkaline and saline alkali soils pH 8-9
3	Lime stone	c	Soils having pH < 6-8
4	Leaching	d	Effective way of saline soils

9.

	A (Amendments)		B (Soil conditions)	
1	Oxygen	a	+380 to + 320 mV E_h	a
2	Nitrogen, Manganese	b	+ 280 to +220 mV E_h	b
3	Iron	c	+ 180 to +150 mV E_h	c
4	Sulphur	d	-120 to -180 mV E_h	d
5	Carbondioxide	e	-200 to -280 mV E_h	e

FORMULAS

1. $Kw = [H^+] [OH^-] = 10^{-14}$

2. $pH = \log 1/[H^+] = -\log [H^+]$

3. $\text{Percentage base saturation} = \dfrac{\text{Exchangeable bases } (\text{cmol}/\text{kg}) \times 100}{\text{CEC (cmol/kg)}}$

4. Lime potential $(LP) = pH - 1/2\ pCa$

5. $\text{Ratio Law} = \dfrac{[H^+]}{\sqrt{[Ca^{+2}]}} = pH - 1/2\ pCa = \text{Constant}$

6. $\text{Gypsum requirement} = \dfrac{(\text{ESP initial} - \text{ESP final}) \times \text{CEC}}{100}$

7. $\text{Phosphate potential} = 1/2\ pCa + P^{H_2PO_4}$

8. $\text{Sodium Adsorption ratio} = SAR = \dfrac{Na^+}{\sqrt{Ca^{2+} + Mg^{2+})/2}}$

9. $pE = -\log (e) = Eh/2.303\ RT\ F^{-1}$

10. $PE = E_o/0.0591$

11. $pE = \dfrac{Eh}{0.0591}$ @ 25°C

12. Mg adsorption ratio $= \dfrac{Mg^{2+}}{Ca^{2+} + Mg^{2+}}$

13. RSC (m el^{-1}) $= (CO_3^{2-} + HCO_3^{-}) - (Ca^{2+} + Mg^{2+})$

14. Chlorine concentration (m el^{-1}) $=$

$$\dfrac{Cl^-}{CO_3^{2-} + HCO_3^{2-} + SO_4^{2-} + Cl^- + NO_3^{2-}}$$

15. Soluble sodium percentage (m el^{-1}) $= \dfrac{Na \times 100}{Ca + Mg + Na}$

KEY TO FILL IN THE BLANKS

1. H$^+$ activity
2. Concentration
3. Sorensen
4. Dissociation constant (Kw)
5. Active acidity
6. Exchangeable Acidity
7. 6.5 -7.5
8. 8.0 to 11.0
9. < 3.5
10. High rainfall
11. Lime
12. Exchangeable H$^+$
13. Exchangeable Al^{+3} and Fe^{+2}
14. N.I=CCE* Fineness factor
15. 10 fold
16. Al, Fe
17. (H$^+$) and (OH$^-$)
18. 90 million hectares
19. Kuttanad area of Kerala in India)
20. EC and ESP
21. Sodium
22. Chlorides and sulphates
23. Carbonates and Bicarbonates
24. Saline soils
25. Alkali soils
26. Salinization and Alkalization processes
27. > 0.1 %
28. Alkali or Sodic soil
29. $D_{dw}/D_{iw} \times 100$
30. $EC_{iw}/EC_{dw} \times 100$
31. 18.6 %
32. 80 percent
33. Na$_2$CO$_3$
34. 0.35:1
35. Salt dilution
36. Osmoregulation
37. Sodic soils
39. Arid climates

40. Osmotic potential
41. reduce oxygen supply
42. Submerged soil
43. Redox potential (Eh)
44. pE
45. Magnesium hazard
46. C1, C2, C3 and C4
47. C1 & C2
48. C3 & C4
49. S1, S2, S3 & S4
50. SAR
51. Salt index
52. negative, positive
53. <1 ppm
54. Puddling
55. Ferrolysis
56. 0.0591
57. NH_4^+ form
58. Methane
59. N, P, Fe, Mn & Mo
60. S, Cu, Zn
61. C_4S_2
62. C_4S_3
63. C_4S_4

KEY TO CHOOSE THE CORRECT ANSWER

1. c	2. d	3. a	4. a	5. a	6. b
7. b	8. a	9. d	10. c	11. a	12. d
13. a	14. b	15. a	16. d	17. b	18. c
19. a	20. a	21. b	22. d	23. b	24. a
25. c	26. b	27. a	28. a	29. a	30. c
31. a	32. c	33. d	34. a	35. b	36. a
37. c	38. d	39. d	40. d	41. e	42. d
43. a	44. a	45. d	46. b	47. a	

KEY TO TRUE OR FALSE

1. True
2. True
3. False [True: It cannot amend the surface acidity]
4. True
5. False [True: Micro nutrients except Mo are available in acidic pH]
6. True

KEY TO MATCH THE FOLLOWING

1.	1-a	2-b	3-C	
2.	1-a	2-c	3-b	
3.	1-c	2-d	3-a	4-e	5-b
4.	1-a	2-c	3-b	4-d
5.	1-b	2-a	3-c	4-d
6.	1-a	2-b	3-c	
7.	1-a	2-c	3-b	4-d
8.	1-a	2-b	3-c	4-d
9.	1-a	2-b	3-c	4-d	5-e

9

WATERSHED MANAGEMENT AND LAND USE PLANNING

M. Raghavendra Reddy and G.S. Yogesh

CHOOSE THE CORRECT ANSWER

1. Soil erosion involves _______________
   - (a) Detachment
   - (b) Transport
   - (c) Deposition
   - (d) All of these

2. Among the following, extent of erosion more in _________
   - (a) Inceptisols
   - (b) Vertisols
   - (c) Alfisols
   - (d) Same for all

3. Which one of the following is not a consequence of soil erosion _______________
   - (a) Land degradation
   - (b) Loss of soil productivity
   - (c) Loss of biodiversity
   - (d) Climate change

4. If the land slope is increased by 4 fold, the velocity of water flowing down the slope becomes approximately _____________
   - (a) Doubled
   - (b) Same rate
   - (c) Half
   - (d) 4 fold

5. If velocity of runoff water is doubled, its energy increases ____________
   - (a) 4 times
   - (b) 2 times
   - (c) 3 times
   - (d) 11/2 times

6. If velocity of run off water is doubled, the quantity of material carries in _________
   - (a) 16 fold
   - (b) 32 fold
   - (c) 8 fold
   - (d) 4 fold

7. Erodability of soil is influenced by its ____________
   - (a) Soil texture
   - (b) Soil structure
   - (c) Nature of clay
   - (d) All of these

8. Soil erosion is influenced by _______________
   - (a) Infiltration capacity
   - (b) Structural stability
   - (c) Soil moisture
   - (d) All of these

9. Lowest rate of soil erosion found in ____________
   - (a) Laterite soils of Tamil Nadu
   - (b) Red soils of Bihar
   - (c) Both a & b
   - (d) Alluvial soils of Gujarat

10. Raindrop has a capability of generating force equal to almost _____ of its weight

    (a) 14 times   (b) 12 times   (c) 10 times   (d) 8 times

11. Splash erosion may take soil away up to __________ metres

    (a) 3   (b) 2   (c) 4   (d) 5

12. Removal of soil by water from small channels ________ erosion

    (a) Splash   (b) Sheet   (c) Rill   (d) Coastal

13. Which one of the following is a special form of erosion

    (a) Stream bank erosion   (b) River bank erosion

    (c) Slumping   (d) All of these

14. Which one of the following is not a form of wind erosion ___________

    (a) Saltation   (b) Suspension   (c) Slumping   (d) Surface creep

15. Saltation carries ___________ sized soil particles

    (a) 0.1 to 0.5 mm   (b) < 0.1 mm

    (c) 0.5 to 1 mm   (d) > 1 mm

16. Suspension carried ___________ sized soil particles

    (a) 0.1 to 0.5 mm   (b) < 0.1 mm

    (c) 0.5 to 1 mm   (d) > 1 mm

17. Saltation accounts __________ % of total wind erosion

    (a) 50 - 75   (b) 40 - 50   (c) 5 - 25   (d) 90

18. Surface creep accounts __________ % of total wind erosion

    (a) 50 - 75   (b) 40 - 50   (c) 5 - 25   (d) 90

19. Rolling and sliding can seen __________ type of erosion

    (a) Saltation   (b) Suspension   (c) Soil creep (d) All of these

20. Clay sized soil particles can be eroded by ____________ process

    (a) Saltation   (b) Suspension   (c) Soil creep (d) All of these

21. Contour bunds are mechanical barriers built _____________ the slope

    (a) Across   (b) Along   (c) Vertically (d) All of these

22. Graded bunds recommended to areas with ___________ % slope

    (a) Up to 10 %   (b) More than 10 %

    (c) Up to 16 %   (d) More than 16 %

23. Which one of the following is not a soil conservation mechanical measure
    _____

    (a) Graded bunds   (b) Bench terrace

    (c) Contour terraces   (d) Contour cultivation

24. Watershed is a __________ unit
    (a) Bio - physical   (b) Socio economical
    (c) Political   (d) All of these

25. ICAR Indian Institute of soil and water conservation headquarters at

    ____

    (a) Ooty   (b) Ballari   (c) Dehradun   (d) Kota

26. Main concept of watershed approach is from __________
    (a) Ridge to valley   (b) Valley to ridge
    (c) Contour to valley   (d) None of these

27. Capability class has no limitation in LCC is __________
    (a) Class I   (b) Class II   (c) Class III   (d) Class IV

28. OC (>1%), CEC (30 C mol (P+) kg^{-1}) & BS (> 80 %) categorized into
    _______ class in LCC

    (a) Clas VIII   (b) Class VI   (c) Class III   (d) Class I

29. Capability index (Ci) - ABCDEFG, where D & G represents
    (a) Gypsum & slope   (b) Depth & Gypsum
    (c) Structure & drainage   (d) None of these

30. Simple management practices is recommended to __________ class of LCC
    (a) Class I   (b) Class II   (c) Class VII   (d) Class VIII

31. Which of the following is not a limitation / Sub classes of LCC
    (a) Erosion   (b) Climate
    (c) Excess water   (d) Salinity

32. LCC is consists of __________
    (a) Capability classes   (b) Capability sub classes
    (c) Capability units   (d) All the above

33. Class of LCC which is very suitable for grazing is __________
    (a) Class I   (b) Class III   (c) Class V   (d) Clas VII

34. In universal soil loss equation, A = RKLSCP : P denotes __________
    (a) Soil conservation practices   (b) Plant height
    (c) Plant geometry   (d) Plant nutrition

35. The process that transforms the productive land to unproductive is called
    as ____

    (a) Forestation   (b) Desertification
    (c) Terracing   (d) Land scaping

FILL IN THE BLANKS

1. Soil erosion exceeds than normal rate called as _______________.

2. Soil fertility is declining in many agricultural lands due to loss of soil matrix in form of _____________.

3. ICAR - NBSS & LUP located at ___________.

4. Soil degradation map available at NBSS & LUP at scale of ________.

5. Expand GLASOD _________________________.

6. Among Alluvial, Black & Red soils, which one will be more affected by erosion _______________.

7. Soil erodability (K) is given by _______________.

8. Removal of soil from the land surface by water is called as _____________.

9. Soil erosion in which soil matrix is lost but remains undetected for a long period called ____________.

10. Advanced stage of rill erosion is _______________.

11. Rills with more than ___________ cm depth are generally called as gullies.

12. ____________ type of special erosion is common in hills.

13. USLE equation given by _______________.

14. $A = R \times K \times L \times S \times$ ___________ $\times P$.

15. Erosion rate per unit of erosion index on cultivated continuous follow on a 9 % slope, 22 m long is ________________.

16. Length of slope influenced to soil loss per unit area according to USLE is ____.

17. Expand USLE ___________________________.

18. Expand RUSLE ___________________________.

19. Wind erosion estimation equation given by ________.

20. Half moon terraces are recommended to fruit trees on steep slopes.

21. Size of sub watershed is _______________.

22. Size of milli watershed is _____________.

23. Size of micro watershed is _______________.

24. Size of mini watershed is ________________.

25. _______________ is defined as natural hydrologic entity, where run off flows to a defined drain or stream.

26. GOI formulated Integrated Watershed Development Programme (IWDP) in _______________ year.

27. Funding pattern of watershed management programme is _______ % by central government and _________ % by state government.

28. _______________ are defined as protection barriers from winds.

29. Lands which are unsuitable for cultivation categorized into _________ class.

30. Number of land capability classes are _________ .

31. Criteria of LCC was given by _______________ .

32. Cultivable classes with no limitations were indicates in LCC by ____ colour.

33. If the land capability class of soil is IIIw, what is its limitation _________ .

34. _______________ classification of soils is to evaluate suitability of mapped soils for sustained use under irrigation.

35. Land suitability for irrigation have _______________ number of classes.

36. Non - arable lands are categorized into _________ class of land suitability for irrigation.

37. If a soil is categorized as S_2 W(2) its limitation is _____________ .

38. Rolling and sweeping of soil particles can seen in _________ type of wind erosion.

39. The process that lowers the current and potential capability of soil to produce goods or services is called _______________ .

40. The type of soil erosion can be easily removed by tillage operation is _________ .

MATCH THE FOLLOWING

1.

A (Soil degradation problem)		B (Extent (%))	
1	Physical deterioration	a	5.5
2	Chemical deterioration	b	4.1
3	Wind erosion	c	45.3
4	Water erosion	d	4.2

2.

A (Land capability class)		B (Extractant)	
1	Class - I	a	Purple
2	Class - II	b	Red
3	Class - VI	c	Orange
4	Class - VII	d	Yellow
5	Class - VIII	e	Green

3.

A (Watershed)		B (Size (ha))	
1	Sub watershed	a	1000 - 10,000
2	Milli watershed	b	10 - 100
3	Micro watershed	c	100 - 1000
4	Mini watershed	d	10,000 - 50,000

4.

A		B	
1	Water erosion	a	Water logging
2	Wind erosion	b	Splash erosion
3	Physical deterioration	c	Salinization
4	Chemical deterioration	d	Saltation

5.

A		B	
1	Saltation	a	< 0.1 mm
2	Suspension	b	0.1 to 0.5 mm
3	Soil creep	c	0.5 to 1 mm

TRUE OR FALSE

1. Water act as nutrient carrier as well as agent of soil erosion.

2. Soil erosion due to run off is more in plains.

3. Geological erosion is normal or natural erosion.

4. Geological erosion is needed to control since it pose big problem.

5. In himalayas, land slides and land slips pose a serious problem.

6. Soil erosion depends on intensity, kinetic energy, amount, duration and frequency of rainfall.

7. When intensity of rainfall is less than in filtration rate, it causes runoff and soil loss.

8. Soil erosion due to loosening and cutting of soil on the river banks is accelerated during flood.

9. Very fine sand behaves more like silt than like sand.

10. Land slides also called as slip erosion.

11. Stream bank erosion is also called as slumping.

12. Bench terraces are flat beds constructed on the hills across the slope.

13. Tap root crops are highly effective in controlling soil erosion.

14. Cowpea and soyabean are erosion resistant crops.

15. Closer spacing of rows across the slope conserve soil.

16. Class IV is non-arable and extremely rocky and only suitable for wild life and recreation

17. Sub classes of LCC indicate both the degree and kind of limitations.

18. Length of growing period is one of the factor of categorize LC(C).

19. Scattering of detached soil particles by impact of rain drops is called splash erosion.

20. In universal soil loss equation, A = RKLSC; K denotes soil erossivity.

KEY TO CHOOSE THE CORRECT ANSWER

1. d	2. c	3. d	4. a	5. a	6. b
7. d	8. d	9. c	10. a	11. b	12. c
13. d	14. c	15. a	16. b	17. a	18. c
19. c	20. b	21. a	22. a	23. d	24. d
25. c	26. a	27. a	28. d	29. a	30. b
31. d	32. d	33. c	34. a	35. b	

KEY TO FILL IN THE BLANKS

1. Accelerated soil erosion
2. Erosion
3. Nagpur
4. 1:250,000
5. Global Assessment of Soil Degradation
6. Red soils
7. Wischmeier and Mannering
8. Water erosion
9. Sheet erosion
10. Gully erosion
11. 30
12. Land slides
13. Weishmeier and Smith

14. C
15. Soil erodability factor
16. 22 m
17. Universal Soil Loss Equation
18. Revised Universal Soil Loss Equation
19. Chepil and Woodruff
20. Fruit & Plantation crops
21. 10,000 - 50,000 ha
22. 1000 - 10,000 ha
23. 100 - 1000 ha
24. 10 - 100 ha
25. Watershed
26. 2008
27. 90 & 10
28. Wind breaks
29. V to VIII
30. 8
31. Klingebiel and Montgometry
32. Green
33. Excess water
34. Land Irrigability Classification
35. Six
36. Class 6
37. Wetness
38. Soil creep
39. Land Degradation
40. Rill erosion

KEY TO MATCH THE FOLLOWING

1.	1-c	2-a	3-b	4-d
2.	1-e	2-d	3-c	4-b	5-a
3.	1-d	2-a	3-c	4-b
4.	1-b	2-d	3-a	4-c
5.	1-b	2-a	3-c

KEY TO TRUE OR FALSE

1. True
2. False [True: Sloping and denuded lands]
3. True
4. False [True: Geological erosion does not pose any problem]
5. True
6. True
7. False [True: More than]
8. True
9. True
10. True
11. False [True: Scour erosion]
12. True
13. False [True: Fibrous root crops]
14. True
15. True
16. False [True: Class VIII]
17. True
18. True
19. True
20. False [True: Soil erodability]

<table><tr><td>

10</td><td>

REMOTE SENSING AND GIS

A. Sathish and M. Raghavendra Reddy</td></tr></table>

CHOOSE THE CORRECT ANSWER

1. Use of Aerial photography was first made during.
   - (a) First World War
   - (b) Second World War
   - (c) 20th Century
   - (d) All of these

2. Aerial photography is arranged through.
   - (a) SOI
   - (b) ISRO
   - (c) NOAA
   - (d) None

3. National Remote Sensing Centre is located at.
   - (a) Bangalore
   - (b) Hyderabad
   - (c) Dehradun
   - (d) Delhi

4. The angle of Coverage in Normal angle photography is.
   - (a) 90°
   - (b) 60°
   - (c) 180°
   - (d) None

5. Aerial photographs provide the additional information of.
   - (a) Lattitude
   - (b) Longitude
   - (c) Altitude
   - (d) None

6. Visual interpretation requires
   - (a) SOI
   - (b) Computer
   - (c) Software
   - (d) None

7. Normalized Difference Vegetation Index is .
   - (a) NIR+R/ NIR-R
   - (b) (NIR-R)+(NIR+R)
   - (c) NIR-R/ NIR+R
   - (d) None

8. The classes that result from unsupervised classification are.
   - (a) Spectral classes
   - (b) Spatial classes
   - (c) Temporal classes
   - (d) None

9. UTM is (a)
   - (a) Software
   - (b) Time Zone
   - (c) Unit
   - (d) Projection

10. Latitude is represented as.
    - (a) North & South
    - (b) East & West
    - (c) West & North
    - (d) South & East

11. A frame of reference for measuring locations on the surface of the earth is.

    (a) Tone   (b) Co-ordinates   (c) Projection (d) Datum

12. The photo-interpretation elements are.

    (a) Tone   (b) Colour   (c) a&b   (d) None

13. Geo-referencing of satellite image is made using.

    (a) Cadastral map (b) Toposheet   (c) Map   (d) None

14. Polyconic is (a)

    (a) Process   (b) Resolution   (c) Projection (d) None

15. Mapping of topography is best made possible by using

    (a) Optical RS data   (b) Aerial Photograph

    (c) Micro wave data   (d) None

16. Digital interpretation of image is commonly called as

    (a) Image analysis   (b) Image enhancement

    (c) Image study   (d) Image processing

17. All along the river (blue colour) white patches are seen on an image which may be identified as Sand, which photo interpretation element is useful here.

    (a) Tone   (b) Association   (c) Shape   (d) All of these

18. Stitching of two images and making it as one image is called .

    (a) Mosaic   (b) Subsetting

    (c) Geo-adding   (d) None

19. Geo-referencing is also called as.

    (a) Geo-correction   (b) Projection

    (c) Geo-rectification   (d) Delhi

20. The file format of raster image used in Erdas imagine software is.

    (a) Geo-Tiff   (b) Img   (c) doc   (d) None

21. Red tone in satellite image indicates

    (a) Water   (b) Road   (c) Vegetation (d) None

22. Each and every pixel in the image can be classified using

    (a) Manual interpretation   (b) Digital interpretation

    (c) Visual interpretation   (d) None

23. The smallest picture element on the map is

    (a) Restoration   (b) Resolution   (c) Pixel   (d) None

24. Type of Aerial Photograph based on the direction of the camera axis is

    (a) Wide angle   (b) Normal angle

    (c) Super wide angle   (d) All

25. Remote sensing became operational with the launch of
    (a) IRS   (b) INSAT   (c) GSAT   (d) None

26. Aerial photography is obtained using.
    (a) Platform   (b) Scanner
    (c) Satellite   (d) Airborne camera

27. GPS helps in obtaining information on .
    (a) Latitude   (b) Longitude   (c) Altitude   (d) All of these

28. When the satellite image shows red tone in all the seasons, the feature is.
    (a) Evergreen forest   (b) Crop land
    (c) Water   (d) None

29. Data into GIS is captured through
    (a) Scanning   (b) Digitization   (c) Digital data   (d) All of these

30. Krigging is
    (a) Software   (b) Geostatistical method
    (c) Step   (d) None

31. LU/LC represents
    (a) Land utilization & land cover   (b) Land use / land cover
    (c) Land usage/land class   (d) None

32. To work with GPS minimum satellites necessary are.
    (a) Four   (b) Twenty Four   (c) Thirty two   (d) None

33. LISS is a
    (a) Software   (b) Camera   (c) Sensor   (d) None

34. Spectral Reflectance values are high for
    (a) Water   (b) Sand   (c) Soil   (d) Forest

35. The satellite imagery which provides highest resolution data of 0.31cm is
    (a) Worldview 3   (b) Quick bird   (c) RISAT 1   (d) Ikonos

36. Landsat satellite belongs to
    (a) India   (b) France   (c) Russia   (d) USA

37. To cover entire India in one scene (satellite imagery) data of sensor to be used is
    (a) LISS IV   (b) PAN   (c) WiFS   (d) None

38. The remote sensing data helps to understand entire area which is called as
    (a) Birds eye view   (b) Synoptic view
    (c) Synchronous view   (d) None

39. The capability to differentiate the spectral reflectance/emittance between various targets is termed as
    (a) Spatial resolution   (b) Radiometric resolution
    (c) Spectral resolution   (d) None

40. Multispectral image is
    (a) Multi-date image   (b) Multi-time image
    (c) Multi-band image   (d) None

41. One commonly used Microwave image data is obtained through
    (a) SAR   (b) RADAR   (c) PAN   (d) None

42. To join two images the operation to be performed is
    (a) Mosaic   (b) Subsetting   (c) Geo-adding   (d) None

43. Soil moisture is estimated using
    (a) Optical data   (b) Microwave data
    (c) Thermal image data   (d) None

44. Thematic map is nothing but
    (a) Map showing specific information
    (b) Map having many information
    (c) Map having no information
    (d) None

45. Sand appears in FCC as
    (a) White   (b) Black   (c) Blue   (d) None

46. Popularly used database management software is
    (a) Microsoft Excel   (b) Microsoft word
    (c) Microsoft Access   (d) None

47. Reflectance value of black body is
    (a) Zero   (b) One   (c) 255   (d) None

48. The path followed by a satellite is referred to as
    (a) Unit   (b) Circuit   (c) Orbit   (d) Pathway

49. The area imaged by satellite is referred to as
    (a) Imagery   (b) Orbit   (c) Swath   (d) Photograph

50. Cartosat II satellite belongs to
    (a) India   (b) USA   (c) Canada   (d) None

51. Image quality is improved through
    (a) Image correction   (b) Image analysis
    (c) Image enhancement   (d) None

52. GIS makes maps
    (a) Static   (b) Versatile   (c) Dull   (d) Dynamic

53. RISAT is a
    (a) Microwave satellite   (b) Optical RS satellite
    (c) Drone   (d) Sensor

54. Creating a smaller dataset from a larger dataset is achieved through
    (a) Importing   (b) Georeferencing   (c) Mosaic   (d) Subset

55. Text book of Remote Sensing and Geographical Information Systems is authored by
    (a) Anji Reddy, M (b) Lillesand   (c) Kumar, S   (d) None

FILL IN THE BLANKS

1. The word Remote Sensing was coined by_________in 1960's.

2. EMR is abbreviated form of _____________________.

3. Before venturing on an operational Remote Sensing Satellite system like IRS, ISRO successfully launched_______________ satellite in 1975.

4. 0.4 to 0.7 mm wavelength region is called ______________.

5. _________is one of the types of atmospheric scatter normally observed.

6. The observations made over the same area on different dates to monitor the objects is called ____________________resolution.

7. SLAR is abbreviated form of___________.

8. _________________is the highest resolution commercial satellite currently available for use in Agriculture.

9. Water bodies appear in _____________colour in a toposheet.

10. LISS is a ____________.

11. Red tone in satellite image indicates _____________________.

12. In case of _______________key, interpreter gradually eliminates all incorrect choices.

13. Each and every pixel in the image can be classified under ____________ interpretation.

14. ___________________ is the smallest picture element on the map.

15. Best time for Aerial Photography is between ____________ and ______________ depending on the meteorological condition prevailing in the area.

16. Types of Aerial Photographs based on the direction of the camera axis are ____________, ______________ and _____________.

17. Image interpretation has various aspects ____________, _________________, _________________, _______________.which have overlapping functions.

18. ______________ resolution is the smallest object that can be detected and distinguished from a point.

19. ______________ colour represents sand in the satellite imagery.

20. GPS space segment consists of _________________ satellites.

21. _______________________ is the software used for small scale database management.

22. _______________________ is considered as father of GIS.

23. _______________________ GIS consists of point, line and polygon features.

24. Erdas imagine is considered as ___________________ GIS software.

25. The usual source of remote sensing data is Electro-magnetic radiation which is ________________ or _________________ from an object.

26. The term "photography" is derived from two Greek words meaning _____________ and ____________________.

27. IRNSS 1E satellite is launched by ______________ in 2016.

28. _______________________ initiated the remote sensing activity in India.

29. The satellite data receiving station is located at ___________________ in India.

30. Electromagnetic radiation is expressed in terms of ____________, _______________ or ___________________.

31. The energy source used in the visible and reflective infrared remote sensing is the ___________________.

32. The visible portion of EMR is from ________to________ in wavelength.

33. The reflectance value for black body is _______ where as for white body it is _____.

34. Geostationary satellites are placed in the orbit at an height of __________ from earth.

35. The path followed by a satellite is referred to as its __________.

36. The area imaged by satellite is referred to as the __________.

37. The remote sensing data is available in __________, __________ and __________ data forms.

38. Black tone in satellite image indicates __________.

39. Use of Aerial photography was first made during __________.

40. NRSC is an acronym of __________ .

41. SOI is __________.

42. Types of Aerial Photographs based on the angle of coverage are __________, __________ and __________.

43. Image interpretation deals with __________, __________, and __________.

44. Cloud in a satellite image is identified with __________ photo interpretation element.

45. __________, __________ and __________ are the three basic steps in supervised classification.

46. __________, __________ and __________ are the general types of errors in GIS database.

47. __________ and __________ are types of GIS data.

48. Digital interpretation is popularly known as __________.

49. SLUSI is an acronym of __________.

50. NBSS&LUP is an acronym of __________.

51. Columns are called __________ and rows are called __________ in DBMS.

52. Map composition is made using __________ software.

53. Point information is extended to polygon information through __________ technique.

54. The commonly used file format in Erdas Imagine software is __________.

55. DGPS is an acronym of __________.

56. One Scientist who is responsible for remote sensing activities in India __________.

57. White tone in satellite image indicates ___________________.

58. Use of Aerial photography was first made during ___________________.

59. In case of ______________ key, interpreter gradually eliminates all incorrect choices.

60. IRS is an acronym of ______________________ .

61. Software used for georeferencing of image is ___________________.

62. NDVI is an acronym of __________.

63. The components of GIS are__________, _________, ____________, ___________ and ___________.

64. Base map includes permanent features like _________, __________, _________, _____________________.

65. WiFS is an acronym of ______________________.

66. The remote sensing data helps to understand entire area which is called as _________________

67. GNSS is an acronym of ______________

68. RADAR is an acronym of __________

69. LiDAR is an acronym of ______________.

70. At _________km, __________km and ___________km respectively from earth, Earth resource satellites, geostationary satellites and GPS satellites are locate(d)

71. Remote sensing was initiated in 1970 to study root-wilt disease of coconut plantations in _________________.

72. Generally satellites are classified as ______ and ____.

73. INSAT 3E is __________________ satellite, where as ____________ is sunsynchronous satellites.

74. The surface directly below the satellite is called the _________________.

75. FCC is an acronym of ________________.

76. Multispectral image is nothing but image having _________.

77. _________________________ is one open source GIS software.

78. Cloud is identified on satellite image using _________________ element.

79. Expand DEM ___________.

80. Expand DGPS ____________.

TRUE OR FALSE

1. Remote Sensing is sensing from a distance with physical contact with the object.

2. Synthetic aperture radar is imaging type radar.

3. Panchromatic data is obtained in three bands.

4. The energy source used in the visible and reflective infrared remote sensing is the sun.

5. The scale 1:50,000 is commonly called as linear scale.

6. Microwave (1 mm - 1 m) region of EMR is used for getting data during cloud cover also.

7. Regions of the electromagnetic spectrum in which the atmosphere is transparent are called atmospheric scatter.

8. SOI toposheet will be used for generating contour information.

9. True colour composite is the commonly used format for analysis.

10. Erdas Imagine is one of the remote sensing software used for image interpretation.

11. Aerial Photograph is best suited for land use mapping.

12. Visual interpretation is better for soil mapping.

13. Digital interpretation uses several band data for analysis.

14. Subsetting of image is done to avoid large size consumption by the file.

15. The interpreter needs the knowledge of the study area well in advance for interpretation.

16. Remote Sensing data is not economical if we want data of inaccessible area.

17. Remote Sensing data has both spatial as well as non spatial information.

18. Image Enhancement deals with manipulation of data for improving its quality for interpretation.

19. Visual interpretation is not good for small area mapping when compared to digital interpretation.

20. The current most widely used vegetation index is the Normalised Detection Vegetation Index.

21. Remote Sensing is observing from far off place with physical contact with the object.

22. IRS is nothing but Indian Remote Sensing.

23. Georeferencing is a method to differentiate two objects.

24. 1:50,000 scale image is having higher information than 1:10,000 scale image.

25. Commonly used projection is UTM.

26. Clipping of image (area of interest) is called as mosaic.

27. File format used in erdas imagine software is .xml.

28. FCC is called as false colour composite.

29. While preparing FCC, minimum three band data is used.

30. Erdas Imagine is a raster GIS software.

31. Supervised classification is not preferred over unsupervised classification.

32. Aerial Photography can be used when stereoscopic analysis is required.

33. In order to prepare soil map of taluk, large scale mapping has to be undertaken.

34. In India the satellite data reception centre is located at Hyderabad.

35. The computer has to be trained by giving training sites before supervised classification.

36. GIS is a tool having both spatial as well as non spatial data.

37. DBMS is an acronym of Database Management Sensor.

38. Aerial photograph can be obtained by any civilian.

39. Photo interpretation elements are required for visual analysis.

40. Aerial photograph provides a current pictorial view of the ground that no map can equal.

41. ESRI is nothing but Environmental Software Research Institute.

42. Maps are static but GIS makes maps dynamic.

43. Raster image is made of pixels in the form of grids.

44. Point , Line and Polygon are the three features in vector GIS.

45. Base map is not required for land use / land cover mapping.

MATCH THE FOLLOWING

1.

	A			B
1	Telescope	A		Nadar
2	RS in India	B		France
3	first Space Station	C		Galileo
4	first stereo-image satellite	D		Dr. Vikaram Sarabhai
5	First aerial photographer	E		Russia

2.

	A		B
1	ISRO	A	Hyderabad
2	NRSC	B	Sriharikota
3	NBSS&LUP	C	Dehradun
4	IIRS	D	Banglore
5	Satish Dhawan Space Centre	E	Ngapur

3.

	A		B
1	Passive RS	A	dust
2	Active RS	B	Smoke
3	Rayleigh scattering	C	Air molecules
4	Mie scattering	D	RADAR
5	Nonselective	E	Sun

4.

	A (Features)		B (FCC)
1	Trees and bushes	A	Dark red
2	Wetland vegetation	B	Blue to grey
3	Water	C	Red
4	Urban areas	D	Blue to black

5.

	A (sensors)		B (spatial resolution)
1	LISS-I	A	56
2	LISS-II	B	36.25
3	Advanced WiFS	C	169-188
4	WiFS	D	5.8
5	PAN	E	72.5

KEY TO CHOOSE THE CORRECT ANSWER

1. a	2. a	3. b	4. b	5. c	6. a
7. c	8. a	9. d	10. a	11. c	12. c
13. b	14. c	15. b	16. d	17. b	18. a
19. c	20. b	21. c	22. b	23. c	24. d
25. a	26. d	27. d	28. a	29. d	30. b
31. b	32. a	33. c	34. b	35. a	36. d
37. c	38. b	39. b	40. c	41. a	42. a

43.	b	44.	a	45.	a	46.	c	47.	a	48.	c
49.	c	50.	a	51.	c	52.	d	53.	a	54.	d
55.	a										

KEY TO FILL IN THE BLANKS

1. Evelyn J Pruit
2. Electromagnetic radiation
3. Aryabhata
4. Visible region
5. Mie
6. Multitemporal
7. Side looking airborne radar
8. World view 3
9. Blue
10. Sensor
11. Vegetation
12. Elimination
13. Digital
14. Pixel
15. 8am to 10 am and 2 to 4 pm
16. Oblique, vertical and horizontal
17. Detection, recognition and identification, analysis, classification and idealization
18. Spatial
19. White
20. 24
21. Microsoft Access
22. Roger Tomlinson
23. Vector
24. Image processing
25. Emitted or reflected
26. Light (phos) and writing (graphien)
27. India
28. Dr. Vikram Sarabai
29. Hyderabad
30. Frequency, energy or wavelength
31. Sun
32. 400 to 700nm
33. Zero and one
34. 36,000km
35. Orbit
36. Swath
37. Image data, Analog and Digital data
38. Waterbodies
39. 1960
40. National Remote Sensing Centre
41. Survey of India
42. Normal angle, wide angle and super wide angle
43. Image correction, image enhancement and information extraction
44. Shadow

45. Training, classification and output

46. Positional errors, Boundary errors and Classification errors

47. Vector and raster

48. Image processing

49. Soil and Land Use Survey of India

50. National Bureau of Soil Survey and Land Use Planning

51. Fields and records

52. Arc GIS

53. Interpolation

54. Img

55. Differential Global Position System

56. Dr. Vikram Sarabai

57. Salt or sand

58. World war One

59. Elimination

60. Indian Remote Sensing

61. Arc GIS

62. Normalised Difference Vegetation Index

63. People, hardware, software, data and applications

64. Roads, railway line, waterbodies, settlements

65. Wide Field Sensor

66. Synoptic coverage

67. Global Navigational Satellite System

68. Radio detection and ranging

69. Light Detection and Ranging

70. 600-900 km, 36,000 km and 22,000 km

71. Kerala

72. Natural satellites and Artificial satellites

73. Geostationary satellite, Cartosat 2

74. Nadir point

75. False Colour Composite

76. Multi band data

77. QGIS

78. Shadow

79. Digital Eleveation model

80. Differential global positioning system

KEY TO TRUE OR FALSE

1. False [True: Remote Sensing is sensing from a distance without physical contact with the object]

2. True

3. False [True: Panchromatic data is obtained in black and white]

4. True

5. False [True: The scale 1:50,000 is can't called as linear scale]

6. True

7. False [True: Regions of the electromagnetic spectrum in which the atmosphere is transparent are called atmospheric windows]

8. True

9. False [True: Flase colour composite is the commonly used format for analysis]

10. True

11. False [True: Aerial Photograph is not suited well for land use mapping]

12. True

13. True

14. True

15. True

16. False [True: Remote Sensing data is economical if we want data of inaccessible area]

17. False [True: Remote Sensing data has both spatial as well as temporal information]

18. True

19. False [True: Visual interpretation is good for small area mapping when compared to digital interpretation]

20. False [True: The current most widely used vegetation index is the Normalised difference Vegetation Index]

21. False [True: Remote Sensing is observing from far off place without physical contact with the object]

22. True

23. False [True: Georeferencing is a method of geocooredinating to locations]

24. False [True: 1:50,000 scale image is having lesser information than 1:10,000 scale image]

25. True

26. False [True: Clipping of image (area of interest) is called as subset]

27. True

28. True

29. True

30. True

31. False [True: Supervised classification is preferred over unsupervised classification]

32. True

33. False [True: In order to prepare soil map of taluk, small scale mapping has to be undertaken]

34. True

35. True

36. True

37. False [True: DBMS is an acronym of Database Management system]

38. False [True: Aerial photograph can't be obtained by every civilian]

39. True

40. True

41. False [True: ESRI is nothing but Environmental systems Research Institute]

42. True

43. True

44. True

45. False [True: Base map is required for land use / land cover mapping]

KEY TO MATCH THE FOLLOWING

1.	1-c	2-d	3-e	4-b	5-a
2.	1-d	2-a	3-e	4-c	5-b
3.	1-e	2-d	3-c	4-b	5-a
4.	1-c	2-a	3-d	4-b	
5.	1-e	2-b	3-a	4-c	5-d

11 RADIOACTIVITY, ISOTOPES AND NANOTECHNOLOGY

Ramya S.H and P.N. Siva Prasad

CHOOSE THE CORRECT ANSWER

1. Radioactive substances emit
   - (a) Hydrogen
   - (b) Nitrogen
   - (c) Helium
   - (d) α, β and γ-rays

2. Natural radioactivity was discovered by
   - (a) Rutherford
   - (b) Madam Curie
   - (c) Henri Becquerel
   - (d) Schmidt

3. Which of the following has the greatest penetrating power?
   - (a) Alpha rays
   - (b) Gamma rays
   - (c) Beta rays
   - (d) None

4. In a beta ray change
   - (a) An electron from the extra-nuclear electrons of the atom is given out
   - (b) An electron from the nucleus is given out
   - (c) A proton from the nucleus is given out
   - (d) A neutron from the nucleus is given out

5. When a radioactive substances is subjected to a vacuum, rate of disintegration per second
   - (a) Increases considerably
   - (b) Increases only if the products are gaseous
   - (c) Is not affected
   - (d) Suffers a slight decrease

6. The half life of radium is 1600 years. After how much time it will remain 25 % of the present?
   - (a) 8000 years
   - (b) 3200 years
   - (c) 1600 years
   - (d) 4800 years

7. When the quantum of radioactive substance is increased two times, the number of atoms disintegrating per unit time is
   (a) Doubled
   (b) Increased but not a great extent
   (c) Increased by square of two
   (d) Not affected

8. A radioactive atom emits a beta ray (an electron) from
   (a) Its outermost orbit
   (b) Its innermost orbit
   (c) Its nucleus
   (d) Not affected

9. The emission of beta rays produces an
   (a) Isotope
   (b) Isobar
   (c) Isotone
   (d) All

10. Which of the following is the isotone of $16\ ^{S32}$?
    (a) $_{15}P^{31}$
    (b) $_{8}O^{16}$
    (c) $_{17}Cl^{35}$
    (d) $_{6}C^{12}$

11. Alpha rays are
    (a) He^{+} ions
    (b) He atoms
    (c) He^{+} ions
    (d) Electrons

12. Isotope differs in
    (a) The number of protons
    (b) The chemical activity
    (c) The valence electrons
    (d) The number of neutrons

13. Elements having different nuclear charge but the same atomic mass are called as
    (a) Isotope
    (b) Isobars
    (c) Isotone
    (d) Isomers

14. The element which is naturally radioactive is
    (a) Nitrogen
    (b) Oxygen
    (c) Phosphorus
    (d) Radium

15. Atomic number and atomic mass of Uranium is 92 and 238. During the course of radioactive disintegration it loses 5 α - particles and 4 β-particles. The atomic number and atomic weight of the new element will be
    (a) At.no.= 90, at.wt.= 233
    (b) At.no.= 86, at.wt.= 218
    (c) At.no.= 82, at.wt.= 218
    (d) At.no.= 88, at.wt.= 228

16. Atomic weight of thorium is 232 and the atomic number is 90. The end product of its disintegration is an isotope of lead (At. Wt. = 208, At.no. = 82). The number of α and β-particles emitted is
    (a) $\alpha = 3; \beta = 3$
    (b) $\alpha = 6; \beta = 0$
    (c) $\alpha = 6; \beta = 4$
    (d) $\alpha = 4; \beta = 6$

17. After 2 hours the amount of radioactive substances is reduced to 1/16 of its original amount. The half –life of the radioactive substances is

(a) 15 minutes

(b) 45 minutes

(c) 30 minutes

(d) 60 minutes

18. A beta ray change results in the production of

(a) An isotope

(b) An allotrope

(c) An isobar

(d) An isomer

19. Artificial radioactive was discovered by

(a) Madame Curie

(b) Irene Curie and Federic Joliot

(c) Fermi

(d) Rutherford

20. If radium and chlorine combines to form radium chloride, the compound is

(a) No longer radioactive

(b) Half as radioactive as radium content

(c) As radioactive as the radium content

(d) Twice as radioactive as the radium content

21. Which of the following statements about radioactivity of an element is incorrect ?

(a) It is a nuclear property

(b) It does not involve any rearrangement of element

(c) Its rate affected by change in temperature and pressure

(d) It remains unaffected by the presence of other element chemically combined with it

22. Atomic weight of carbon, nitrogen and oxygen are 12, 14 and 16 respectively. An atom of atomic weight and nuclear charge +6 is an isotope of

(a) Oxygen

(b) Carbon

(c) Nitrogen

(d) None

23. Existence of isotope was first established by

(a) E.Rutherford

(b) F.W. Aston

(c) J.Chadwick

(d) W.Rontgen

24. Other isotope of chlorine will have the same

(a) Mass number

(b) Neutron content

(c) Atomic number

(d) None

25. The phenomenon of radioactivity is associated with

(a) Emission of electrons only

(b) Emission of protons only

(c) Emission of helium ions

(d) Decay of nuclei

26. Isotopes differ in

(a) The number of protons

(b) The valence number

(c) The chemical activity

(d) The number of neutrons

27. Which of the following statement is incorrect?

    (a) Isobars possess the same chemical properties

    (b) Isotopes occupy the same position in the periodic table

    (c) In isobars the total number of protons and neutrons in the nucleus is same

    (d) Isotopes possess same atomic number

28. Isotopes of an element have the same

    (a) Atomic number   (b) Atomic weight

    (c) Number of neutrons in their nucleus (d) Atomic volume

29. $_{17}Cl^{35}$ and $_{17}Cl^{37}$ are

    (a) Isotopes   (b) Isotones   (c) Isomers   (d) Isobars

30. Loss of one α-ray followed by two β-rays from a radio element results in the formation of an

    (a) Isomer   (b) Isotone   (c) Isobar   (d) Isotope

31. The nucleus of U^{234} and U^{238} contain

    (a) The same number of neutrons

    (b) The same number of protons

    (c) Different number of protons

    (d) None

32. Atomic nuclei containing the same number of protons but different number of neutrons are called

    (a) Isobar   (b) Isotopes   (c) Isotone   (d) Isomer

33. The emission of a á-ray from an atom of $90Th^{234}$ results in the formation of its

    (a) Isotope   (b) Isobar   (c) Isotone   (d) Isomer

34. $_{14}S^{30}$, $_{15}P^{31}$, $_{16}S^{32}$ are

    (a) Isotopes   (b) Isotones   (c) Isomers   (d) Isobars

35. If the atom of uranium of mass number 238 and atomic number 92 absorbs a neutron and then breaks up emitting an alpha particle, the mass number of the new element will be

    (a) 236   (b) 235   (c) 234   (d) Unchanged

36. $_{1}H^{1}$, $_{1}H^{2}$, $_{1}H^{3}$ are

    (a) Isotopes   (b) Isobars   (c) Isotones   (d) Isomers

37. Radium is a radioactive element.Its electronic configuration shows than it has only two electrons in the outermost orbit and 86 electrons in the inner orbits. The atomic weight of radium is 226.07.The number of neutrons present in the radium nucleus will be

    (a) 140.07　　　　(b) 138　　　　(c) 140　　　　(d) 88

38. The half-life of a radioactive element is 35 years. If there are 4×10^6 nuclei at the start, then after how many years they will be left 0.5×10^6?

    (a) 35　　　　(b) 105　　　　(c) 70　　　　(d) 140

39. If 4 g of a radioactive isotope has half life of 10 days, the half life of 2 g sample is

    (a) 10 days　　　　(b) 30 days　　　　(c) 20 days　　　　(d) 5 days

40. The emission of a á ray produces

    (a) Isobar　　　　(b) Isotone　　　　(c) Isotope　　　　(d) Isomer

41. α–rays consist of a stream of

    (a) H

    (b) He^+ ions

    (c) Only electrons

    (d) Only neutrons

42. Tritium is an isotope of

    (a) Tellurium　　　　(b) Tritanium　　　　(c) Tentalum　　　(d) Hydrogen

43. A device used for the measurement of radioactivity is a

    (a) Mass spectrometer

    (b) Cyclotron

    (c) Nuclear reactor

    (d) G.M.Counter

44. Which of the following is a radioactive inert gas?

    (a) Ne　　　　(b) Rn　　　　(c) Ar　　　　(d) Xe

45. If a radioactive wastes must be stored for seven half-lives before disposal

    (a) How long must $_{15}P^{31}$ ($t_{\frac{1}{2}}$ = 14.3 days) be held?

    (b) What fraction of $_{15}P^{32}$ originally set aside remains after this time?

    (c) 7 half lives of $_{15}P^{32}$ = 7 × 14.3 days =100 days

    (d) Mass remaining = original $/2^7$ = 1/128 of the original amount

46. Nanoparticles can be synthesized by

    (a) Physical

    (b) chemical

    (c) Physico-chemical and biological

    (d) All these methods

47. Physical method of nanoparticle production includes

    (a) GrindingandSputtering

    (b) Thermal evaporation

    (c) Pulser Laser Deposition Technique

    (d) All of these

48. Which method is most efficient in nanoparticle production
    (a) Physical method
    (b) chemical method
    (c) Physico-chemical (Aerosol method)
    (d) All these methods

49. Safe and ecologically sound approach for nanoparticle production
    (a) Physical   (b) Chemical
    (c) Biological   (d) Physico-chemical

FILL IN THE BLANKS

1. If an atom the number of electrons outside the nucleus is equal to the number of ________________.

2. As the value of orbit number increases the size of the atom will __________.

3. Negative ions are formed from neutral atoms by gain of ____________.

4. Elements having different nuclear charge but the same atomic mass are called as ____________.

5. Atomic nuclei containing the same number of protons but different number of neutrons are called ________________.

6. A device used for the measurement of radioactivity is ____________.

7. The heat content of the system is called as ____________.

8. The number molecules present in one gram molecule of a gas is known as ____________.

9. The most electronegative element in periodic table is ____________

10. The element with atomic number 58 to 71 is knows as ____________

11. The transitional elements are known as ____________.

12. The elements are classified in the periodic table according to ________________.

13. The noble gas forming the maximum number of compounds is __________.

14. The process sof passing a precipitate into colloidal solution on adding an electrolyte called ____________.

15. Scattering of light by colloidal particles is called as ____________.

16. __________________ is the art and science of manipulating matter at the nanoscale (1 to 100 nm) to create new and unique materials and products with enormous potential to change society.

17. __________________ is emerging as the sixth revolutionary technology in the current era.

18. 1 nm is __________________.

19. PSA means __________________.

20. Expansion of TEM __________________.

21. Full form of SEM __________________.

22. Abbreviation of AFM __________________.

23. FTIR Means __________________.

24. Expand ICPMS __________________.

25. QCM is the __________________.

26. Founder of nanoscience __________________.

27. The term nanotechnology was coined by __________________.

28. The term nanotechnology was introduced into the world by __________________ during __________________.

29. Carbon nanotubes is mainly used for __________________.

30. Efficient delivery of fertilizers, pesticides and herbicides by __________________.

31. __________________ is the basic material for the development of nanotechnology.

32. Carbons nanotubes were discovered by __________________.

33. The term nano derived from the Greek word nano's which means __________________.

34. Nanoscience applies to the particles who size varies from __________________.

35. Nano materials have an increased __________________ compared to bulk materials.

36. __________________ is considered as a very important phenomenon in predicting the behavior of nanoparticles.

37. As the particle size decreases brownian motion will __________________.

38. Nanoparticles agglomeration occurs when net interaction potential becomes purely attractive at __________________ and __________________ .

39. __________________ and __________________ plays an important role in deciding the behaviour of the nanoparticles.

40. Higher the zeta potential higher will be the stability ____________.

41. Lower zeta potential results in ____________.

42. ____________ and ____________ approaches are considering in generating nanoparticles.

43. Top-down approach was given by ____________.

44. Bottom –up approach was given by ____________.

45. The hydroxyl group of the poly vinyl pyrolidene acts as an appropriate reductant for aqueous synthesis of nanoparticles made of ____________.

46. Formation of nanoparticles by the vapours of the metal is ____________.

47. ____________ is considered as most accurate method for nanoparticle production.

48. ____________ Produces phosphorus nanoparticles from tricalcium phosphate.

49. Fungi is more efficient in synthesis of nanoparticles compared to ____________.

50. *Azadirachta indica* produces ____________ nanoparticles.

51. Proteins (32 and 33 kDA group) can be used for the production of ____________ nanoparticles from their respective oxide salts.

52. Nanoparticles produced by *Desmodium triflorum* have good ____________ Properties against the pathogen.

53. ____________ is the technique of defining useful shapes on the surface of a semiconductor wafer.

54. ____________ is the crucial technique in fabrication of nanoparticles.

55. ____________ detects the nanparticles with extremely low concentration of order 10^{-12} g (pg)

56. ____________ method is mainly used for the determination of gold nanoparticles.

57. ____________ are formed as result of intensive dissolution of primary soil minerals.

58. Nanoparticle with lower zeta potential tend to ____________.

59. If zeta potential of more than ± 61 indicates ____________.

60. ____________ can be used as a insecticide instead of conventional insecticide.

61. An Indian agro-scientist, Dr. J. C Tarafdar has innovated ____________ using biosynthesis, for the first time in the world.

KEY TO CHOOSE THE CORRECT ANSWER

1. d	2. c	3. c	4. a	5. c	6. b
7. a	8. a	9. b	10. a	11. d	12. d
13. b	14. d	15. b	16. b	17. c	18. c
19. a	20. c	21. c	22. b	23. b	24. c
25. d	26. d	27. c	28. a	29. a	30. d
31. b	32. b	33. b	34. b	35. c	36. a
37. b	38. b	39. a	40. a	41. b	42. d
43. d	44. b	45. a	46. d	47. d	48. c
49. c					

KEY TO FILL IN THE BLANKS

1. Protons
2. Increases
3. Electrons
4. Isobars
5. Isotopes
6. G.M. counters
7. Enthalpy
8. Avogadro's number
9. Fluoride
10. Lanthanides
11. d-block elements
12. Atomic number
13. Xenon
14. Dialysis
15. Tyndal effect
16. Nanotechnology
17. Nanotechnology
18. 1 billionth (10^{-9}) of a meter
19. Particle Size Analyzer
20. Transmission Electron Microscopy
21. Scanning Electron Microscope
22. Atomic Force Microscopy
23. Fourier Transform Infrared Spectroscopy
24. Inductively Coupled Plasma Mass Spectrometry
25. Quartz Crystal Microgravimetry
26. Nobel Laureate Richard Feynman
27. Norio Taniguchi
28. Eric Drexler , 1986
29. Seed germination
30. Nananoscale carrier
31. Fullerenes
32. S. Lijima in 1991
33. Dwarf
34. 1 to 100 nm

35. Surface to volume ratio
36. Brownian Motion
37. Increases
38. Long and short range
39. Zeta potential and pH
40. Nanoparticles
41. Coagulation
42. Top-down and bottom-up
43. Richard Feynman
44. Jean – Marie Lehn
45. Palladium, Platinum, silver and gold
46. Homogeneous nucleation
47. Electrospray
48. *Bacillus megaterium* JCT 13
49. Bacteria
50. Silver
51. Mg, Zn and Fe
52. Antibacterial
53. Lithography
54. Lithography
55. ICP-MS
56. ICP-MS
57. Nanorods
58. Coagulation /Flocculation
59. Excellent stability
60. Surface modified hydrophobic nanosilica
61. Nano-fertilisers

12   AGRICULTURAL STATISTICS

V.R.R. Parama and M. Raghavendra Reddy

CHOOSE THE CORRECT ANSWER

1. Statistics deals with _______________
   - (a) Qualitative information
   - (b) Quantitative information
   - (c) Both a & b
   - (d) None of these

2. Which of the following represents data ?
   - (a) a single value
   - (b) Only two values in a set
   - (c) a group of values in a set
   - (d) None of these

3. Data taken from the publication "NBSS & LUP (2010)" will be considered as ____
   - (a) Primary data
   - (b) Secondary data
   - (c) Primary and Secondary data
   - (d) Neither primary nor secondary data

4. Primary data is _______________
   - (a) More reliable and accurate
   - (b) Only accurate
   - (c) Only reliable
   - (d) Neither primary nor secondary data

5. Class interval is measured as _______________
   - (a) The sum of the upper and lower limit
   - (b) Half of the sum of lower and upper limit
   - (c) Half of the difference between upper and lower limit
   - (d) The difference between upper and lower limit

6. Histogram is suitable for the data presented as _______________
   - (a) Continuous grouped frequency distribution
   - (b) Discrete grouped frequency distribution
   - (c) Individual series
   - (d) All the above

7. Which of the following is a discrete variables _______________
   (a) Height of plants   (b) Length of leaves
   (c) Number of spikes   (d) Age of plants

8. Which of the following is an attribute _______________
   (a) Height of persons   (b) Colour of eyes
   (c) Temperature   (d) Amount of rainfall

9. The most ideal average is _______________
   (a) Median   (b) Mean
   (c) Mode   (d) None of these

10. The average which is affected by extreme values is _______________
   (a) Mean   (b) Median   (c) Mode   (d) Quartiles

11. The average which is located through histogram is _______________
   (a) Mean   (b) Median   (c) Mode   (d) Quartiles

12. Among the following which is not a measure of central tendency _______________
   (a) Mean   (b) Median   (c) Mode   (d) Range

13. The value which divides the series into two equal parts is _______________
   (a) Mean   (b) Median   (c) Mode   (d) HM

14. The value which occurs more frequently is called _______________
   (a) Mean   (b) Median   (c) Mode   (d) GM

15. The geometric mean of 2, 4 and 0 is _______________
   (a) 8   (b) 8   (c) 0   (d) 3

16. The computed average is _______________
   (a) Mean   (b) Median   (c) Mode   (d) Quartiles

17. The correct relationship between AM, GM and HM is _______________
   (a) AM = GM = HM   (b) GM > AM > HM
   (c) HM > GM > AM   (d) AM > GM > HM

18. Harmonic mean gives more weightage to _______________
   (a) Small values   (b) Large values
   (c) Positive values   (d) Negative values

19. In a distribution the value around which the items tend to be most heavily
    concentrated is called _______________
   (a) Mean   (b) Median
   (c) Third Quartile   (d) Mode

20. The suitable measure of central tendency for qualitative data is ____________

    (a) Mode   (b) Arithmetic mean

    (c) Geometric mean   (d) Median

21. The partition value which divide a series into two equal parts is known as

    ______

    (a) Second quartile   (b) Third quartile

    (c) Fourth quartile   (d) Sixth decile

22. Mode is that value in a frequency distribution which possesses ____________

    (a) Minimum frequency   (b) Maximum frequency

    (c) Frequency one   (d) None of these

23. Soil temperature is a __________________

    (a) Discrete variable

    (b) Both discrete & continuous variable

    (c) Continuous variable

    (d) None of these

24. Which measure of dispersion ensures highest degree of reliability _________

    (a) Range   (b) Mean deviation

    (c) Quartile deviation   (d) Standard deviation

25. Which measure of dispersion can be calculated in case of open end class intervals ______________

    (a) Range   (b) Standard deviation

    (c) Co efficient of variation   (d) Quartile deviation

26. If the minimum value in a set is 9 and its range is 57, the maximum value of the set is __________________

    (a) 33   (b) 66   (c) 48   (d) None of these

27. If all values in a sample are same, then their variance is ______________

    (a) Zero   (b) One

    (c) Not calculable   (d) All of these

28. If in an experimental plot, all the plants have the same height then the variance is

    (a) Constant   (b) Zero   (c) One   (d) None of these

29. The standard deviation of a set of values will be __________________

    (a) Positive when the values are positive

    (b) Positive when the values are negative

    (c) Always positive   (d) All of these

30. A measure which gives more weight to extreme values and less to those which are near the mean is ________________

    (a) Mean deviation
    (b) Standard deviation
    (c) Range
    (d) No measures

31. For a positive skewed distribution, the relation between mean, median and mode is ________________

    (a) Median > Mean > Mode
    (b) Mean > Median > Mode
    (c) Mean = Median = Mode
    (d) Mean < Median < Mode

32. The unit of correlation coefficient is ________________

    (a) Kg/cc
    (b) Percent
    (c) Unit less
    (d) None of these

33. The range of simple correlation coefficient is ________________

    (a) 0 to ∞
    (b) $-\infty$ to ∞
    (c) 0 to 1
    (d) -1 to +1

34. When two variables vary in opposite directions then correlation coefficient will be ________________

    (a) Positive
    (b) Negative
    (c) Zero
    (d) None of these

35. When covariance between X and Y is zero then correlation coefficient is ________

    (a) 1
    (b) -1
    (c) 0
    (d) $\pm$ 1

36. Cause and effect relationship can not be revealed by ________________

    (a) Regression
    (b) Correlation
    (c) Variance
    (d) Mean

37. Correlation between two variables is called ________________

    (a) Multiple correlation
    (b) Partial correlation
    (c) Regression
    (d) Simple correlation

38. The range of rank correlation coefficient is ________________

    (a) 0 to 1
    (b) -1 to 1
    (c) 0 to ∞
    (d) always zero

39. The term regression was introduced by ________________

    (a) R.A. Fisher
    (b) Sir Francis Galton
    (c) Karl Pearson
    (d) None of these

40. Regression equation is also named as ________________

    (a) Prediction equation
    (b) Estimating equation
    (c) Line of average relationship
    (d) All of these

41. If $\rho = \pm 1$, the two lines of regressions are ________________

    (a) Coincident
    (b) Parallel
    (c) Perpendicular to each other
    (d) None of these

42. If $\rho \pm 1$, the angle between the two lines of regression is ___________

    (a) Zero degree   (b) Ninety degree

    (c) Sixty degree   (d) Thirty degree

43. If $\rho = 0$, the lines of regression are ___________

    (a) Coincident   (b) Parallel

    (c) Perpendicular to each other   (d) None of these

44. If $\rho = 0$, the angle between the two lines of regression is

    (a) Zero degree   (b) Ninety degree

    (c) Sixty degree   (d) Thirty degree

45. Regression co efficient is independent of ___________

    (a) Origin   (b) Scale

    (c) Both origin and scale   (d) Neither origin nor scale

46. The total area under the normal curve is ___________

    (a) 0   (b) 1   (c) ρ (d) None

47. Normal distribution is a

    (a) Discrete distribution

    (b) Discrete & continuous distribution

    (c) Continuous distribution   (d) None of the above

48. The normal distribution is

    (a) Symmetric about $x = \mu$ (b) Asymmetric about $x = \mu$

    (c) Skewed to the right   (d) Skewed to the left

49. The part of population is called

    (a) Population   (b) Sample   (c) Universe   (d) None

50. The characteristics of the population is called

    (a) Parameter   (b) Statistic   (c) Population (d) Sample

51. The characteristics of the sample is called

    (a) Parameter   (b) Statistic   (c) Population (d) Sample

52. The process by which units are selected with equal probability is termed as

    (a) Random sampling   (b) Non random sampling

    (c) Sampling   (d) Complete enumeration

53. The standard deviation of sampling distribution of means is

    (a) Sampling variance of mean   (b) Standard deviation of mean

    (c) Standard error of mean   (d) None

54. The standard error of mean is given by

    (a) σ　　　(b) σ^2 / n　　　(c) $\sigma / \sqrt{n}$　　　(d) σ^2 / n^2

55. In an agricultural experiments, the variables such as fertilizers, spacing, irrigation Schedule, pesticides etc, are known as ____________________
    (a) Responses　　　　　　　　　　(b) Factors
    (c) Both　　　　　　　　　　　　(d) None of these

56. Experimental units are ____________________
    (a) Basic objects of experiments　　(b) Factors
    (c) Treatments　　　　　　　　　　(d) Replications

57. The variations in responses caused by the extraneous —— factors is termed as ______
    (a) Environmental error　　　　　　(b) Human error
    (c) Design error　　　　　　　　　(d) Experimental error

58. Error degrees of freedom of a experiment is should not be less than ________
    (a) 12　　　　　(b) 10　　　　　(c) 8　　　　　(d) 14

59. When all the treatments have equal chances of being allocated to different Experimental units is known as ______________
    (a) Replications　　　　　　　　　(b) Randomization
    (c) Local control　　　　　　　　　(d) Treatments

60. Grouping of homogeneous experimental units into blocks is ______________
    (a) Blocking　　　　　　　　　　(b) Imposition of treatments
    (c) Local control of error　　　　　(d) Randomization

61. Determination of optimum plot size is done by ______________
    (a) Maximum curvature method　　(b) Fair field smiths variance law
    (c) Both　　　　　　　　　　　　(d) None of these

62. Number of factors in completely randomised design is ______________
    (a) 4　　　　　(b) 3　　　　　(c) 2　　　　　(d) 1

63. Error degrees of freedom for CRD is ______________
    (a) $(n-t)$　　　(b) $(t-1)$　　　(c) $(rt-1)$　　　(d) $(r-1)(t-1)$

64. Error degrees of freedom for RBD is ______________
    (a) $(r-1)$　　　(b) $(t-1)$　　　(c) $(rt-1)$　　　(d) $(r-1)(t-1)$

65. Number of treatments and number of replications are equal in ____________
    (a) RBD　　　　(b) LSD　　　　(c) CRD　　　　(d) All of these

66. Maximum number of treatments that can imposed in LSD is ___________
    (a) 8   (b) 10   (c) 12   (d) 14

67. Minimum number of treatments that can be imposed in LSD is __________
    (a) 7   (b) 6   (c) 8   (d) 5

68. The experimental units will be squared exactly in ______________ design
    (a) CRD   (b) RBD   (c) LSD   (d) Split

69. Error degrees of freedom for LSD is ________________
    (a) (t-1) (t-1)   (b) (t-1) (t-2)
    (c) (t-2) (t-2)   (d) (t-3) (t-1)

70. Combined effect of 2 or more factors when they are used simultaneously can seen in ____________ experiments
    (a) Factorial   (b) CRD   (c) RBD   (d) LSD

71. The term ______________ refers to a set of related treatments
    (a) Levels of a factor   (b) Factor
    (c) Treatments   (d) Replications

72. Number of factor in 2^4 factorial experiments is ________________
    (a) 2   (b) 24   (c) 8   (d) 4

73. ______________ design helps in saving the experimental material
    (a) Split plot   (b) CRC   (c) LSD   (d) RBD

74. A split plot having three main treatments, 3 replications with 4 sub plots has an error degrees of freedom is ________________
    (a) 12   (b) 18   (c) 15   (d) 14

FILL IN THE BLANKS

1.   Statistics is a ____________________________.

2.   Statistics deals with only _________________________.

3.   Statistics does not study ________________.

4.   __________________ is the first stage in any statistical investigation.

5.   Classification is the ____________ of related facts.

6.   A data collected for the first time is called _________________.

7.   A data collected from published sources is called ____________.

8.   Direct personal interview is one of the methods of __________ data collection.

9. Classification of data according to attributes is known as ______________ classification.

10. Classification based on measurable characteristics is known as ______________ classification.

11. The difference between the upper and lower limit of a class ______________.

12. The difference between largest value and smallest value in a data is called ______________.

13. ______________ is the most ideal measure of central tendency.

14. The mean of the values 11, 12, 13, 14 and 15 is ______________.

15. ______________ cannot be represented graphically.

16. Median is a ______________ average.

17. ______________ is a middle most value in a ordered series of data.

18. A central value which divides the series into two equal parts is called ______________.

19. The median of the series 3, 18, 7, 20, 11, 12, 9, 17, 22 is ______________.

20. ______________ is a value which occurs more frequently in a series.

21. The mode of the distribution of values 5, 7, 9, 9, 8, 5, 6, 8, 7, 7, 5, 7, 9, 2, 7 is ______________.

22. The geometric mean of 2, 4, 8 is ______________.

23. The reciprocal of the arithmetic mean of the reciprocals of individual observations is called as ______________.

24. The relationship between AM, GM, HM is ______________.

25. Range is defined as the difference between ______________ and ______________ Values.

26. The best measure of dispersion is ______________.

27. Root mean square deviation is also termed as ______________.

28. Standard deviation is the ______________ of variance.

29. Co efficient of variation is defined as ______________.

30. Lesser the ______________ the more homogeneous the data.

31. High value of coefficient of variation indicates ______________ consistency in data.

32. Standard deviation gives more weight to ______________ values.

33. Mean deviation from median is always ______________ than mean deviation from mean.

34. ________________ represents lack of symmetry of frequency distribution.

35. For a symmetric distribution coefficient of skewness is ______.

36. In case of normal distribution, mean, median and mode are _______.

37. Kurtosis means _____________ of frequency curve.

38. ________________ is a study if relationship between two variables.

39. Correlation between any two variables is called _______________.

40. The measure of correlation is always ___________.

41. When two variables vary in the same direction then correlation is said to be _____________.

42. The range of correlation coefficient is ___________. (-1 to +1)

43. When r = 1, then correlation is said to be _________________________.

44. ________________ analysis is meant to study functional relationship between two Variables.

45. A regression model is a _______________ relationship between the dependent and independent variables.

46. The measure of change in dependent variable corresponding to an unit change in independent variable is called ___________________________.

47. If the variables X and Y are independent, the value of regression coefficient is ________.

48. The regression reveals cause and effect relationship.

49. Regression coefficient ranges from _________ to ____________.

50. The shape of the normal curve is _______________.

51. The normal distribution is a ____________ probability distribution.

52. The values if mean, median and mode are equal in _________________.

53. The totality of individuals under a given phenomenon is termed as ____________.

54. The part of the population of interest is called __________.

55. The process of drawing a sample from a population is termed as ___________.

56. The characteristics of a population is termed as _______________.

57. The characteristics of a sample of observation is termed as ____________.

58. The _______________ of the population are unknown constants.

59. The _______________ obtained from the sample are random variables.

60. The probability sample is also termed as ______________.

61. Modern concepts of experimental design are given by ______________.

62. Repeated application of the treatments of experiments is known as ______________.

63. Sampling error is the difference between the entire plot value and the ______________ Value.

64. Fundamental principles of ANOVA were developed by______________.

65. Most commonly used transformations are ______________, ______________ and ______________ transformations.

66. Missing values are estimated using the method developed by______________.

67. CRD is appropriate only when experimental material is ______________.

68. Unequal number of replications can be seen in ______________ design.

69. The precision of experiment in RBD is increases by a —— of ______________.

70. If an experiment is designed with randomized blocks with 3 replications and 8 Treatments, the error degrees of freedom of design is ______________.

71. Minimum error degrees of freedom needed in a design conduct an experiment for agriculture is ______________.

72. Number of levels in a 3^4 factorial experiment is ______________.

73. ______________ sampling is used for selecting the test farms.

MATCH THE FOLLOWING

1.

	A		B
1	Mean	a	Number of observation falling into each class
2	Median	b	Can be located through histogram
3	Frequency	c	Depends on all the observations
4	Geometric mean	d	Can be located through ogive
5	Quartiles	e	Two dimensional diagram
6	Mode	f	Divides the series into four equal parts
7	Histogram	g	Used to find rates of change

2.

	A			B
1	Mean		a	Root mean square deviation
2	Median		b	Ideal measure of central tendency
3	Variance		c	Used to compute rates & ratios
4	Range		d	Positional average
5	Coefficient of Variation		e	Mean square deviation
6	Standard deviation		f	Difference between highest & lowest values
7	Geometric mean		g	Measure of consistency

3.

	A (Design)			B (edf)
1	CRD		a	(t-1) (t-2)
2	RCBD		b	m (r-1) (s-1)
3	LSD		c	(r-1) (t-1)
4	Split Split plot design		d	(n-t)
5	Split Plot design		e	Ab (r-1) (c-1)

TRUE OR FALSE

1. Secondary data is more reliable and accurate than primary data.

2. The number of observations falling into a particular class is known as class frequency.

3. Qualitative facts cannot be presented in the form of a table.

4. An organized data is called raw data.

5. Measure of dispersion suitable for comparing any two series is coefficient of variation.

6. Variance can never be negative.

7. For calculating variance, deviations are taken from mean.

8. If variance of a set observations is 64 then its S.D is 4.

9. When two variables vary in opposite directions then correlation is said to be negative.

10. Correlation coefficient is dependent of both origin and scale.

11. Two independent variables are uncorrelated.

12. If p denotes the correlation coefficient then p2 is called coefficient of determination.

13. The other name of regression equation is prediction equation.

14. A regression model always linear.

15. The geometric mean of two regression coefficient is correlation coefficient.

16. The regression coefficient is dependent of origin but not scale.

17. The design of an experiment is just like construction of a house.

18. When the information from an experiment is more the precision of the experiment will be less.

19. Local control can be accomplished by uniformity trial.

20. Using fertility contour map homogenous experimental units cant be grouped as blocks.

21. Experimental units should be homogenous.

22. An increased size block may results in variability within the block may be increased and decreased size block may loss the degree of precision.

23. CRD is preferred for field experiments.

24. Blocks and replications are used synonymously in complete block designs.

25. In fertility gradient/ slope are found in one direction (ray from N to S) then the blocks are termed in opposite direction (from east to west).

26. Amount of information can be get more in CRD than RBD.

KEY TO CHOOSE THE CORRECT ANSWER

1. b	2. c	3. b	4. a	5. d	6. a
7. c	8. b	9. b	10. a	11. c	12. d
13. b	14. c	15. c	16. a	17. d	18. a
19. d	20. d	21. a	22. b	23. c	24. d
25. d	26. b	27. a	28. b	29. c	30. b
31. b	32. c	33. d	34. b	35. c	36. b
37. d	38. b	39. b	40. d	41. a	42. a
43. c	44. b	45. a	46. b	47. c	48. a

49. b	50. a	51. b	52. a	53. c	54. c
55. b	56. a	57. d	58. a	59. b	60. c
61. c	62. d	63. a	64. d	65. b	66. c
67. d	68. c	69. b	70. a	71. b	72. d
73. a	74. b				

KEY TO FILL IN THE BLANKS

1. Numerical statement of facts	2. Quantitative characteristics
3. Individuals	4. Collection of data
5. Grouping	6. Primary data
7. Secondary data	8. Primary
9. Qualitative	10. Quantitative
11. Class interval	12. Range
13. Mean	14. 13
15. Mean	16. Positional
17. Median	18. Median
19. 12	20. Mode
21. 7	22. 4
23. Harmonic mean	24. AM > GM > HM
25. Maximum and Minimum	26. Standard deviation
27. Standard deviation	28. Square root
29. the ratio of standard deviation to its mean expressed as a percentage	30. Co efficient of variation
31. less	32. extreme
33. less	34. Skewness
35. Zero	36. equal
37. Peakness	38. Correlation
39. Simple correlation	40. Unit less
41. Positive	42. -1 to +1
43. Perfectly positive or linear	44. Regression
45. Functional	46. regression coefficient

47. Zero
48. regression
49. -1 to +1
50. bell shaped
51. Continuous
52. normal distribution
53. Population
54. sample
55. sampling
56. Parameter
57. statistic
58. Parameters
59. Statistic
60. random sample
61. R.A.Fisher
62. Replication
63. Sample
64. R.A.Fisher
65. Log, Square root and Angular
66. Yates
67. Homogeneous
68. CRD
69. Local control
70. 14
71. 12
72. 3
73. Stratified random

KEY TO MATCH THE FOLLOWING

1.   1-c   2-d   3-a   4-g   5-f
   6-b   7-e
2.   1-b   2-d   3-e   4-f   5-g
   6-a   7-c
3.   1-d   2-c   3-a   4-e   5-b

KEY TO TRUE OR FALSE

1. False [True: Primary data is more reliable and accurate than secondary data]
2. True
3. True
4. False [Tue: An unorganized data is called raw data]
5. True
6. True
7. True
8. False [True: If variance of a set of observations is 64 then its S.D is 8]
9. True
10. False [True: Correlation coefficient is independent of both origin and scale]
11. True

12. True

13. True

14. False [True: A regression model may be linear or non linear]

15. True

16. False [True: The regression coefficient is independent of origin but not scale]

17. True

18. False [True: When the information from an experiment is more, the precision of the experiment will be more]

19. True

20. False [True: Using fertility contour map, homogenous experimental units can be grouped as blocks]

21. True

22. True

23. False [True: CRD is not preferred for field experiments]

24. True

25. True

26. False [Amount of information can be get more in RBD than CRD. Hence RBD is more efficient than CRD]

IMPORTANT FACTS IN SOIL SCIENCE

M. Raghavendra Reddy and
Adiveppa Mallappa Asangi

Size of soil particles

Soil particles	USDA	ISSS
Clay	<0.002 mm	0.002 mm
Silt	0.002 –0.05 mm	0.02 mm
Very fine sand	0.05 –0.1 mm	-
Fine sand	0.1 –0.25 mm	0.2 mm
Medium sand	0.25 –0.5 mm	-
Coarse sand	0.5 –1.0 mm	2 mm
Very coarse	1.0 –2.0 mm	-

Cation exchange capacity (CEC) of soils

Soil texture	$[C\ mol\ (p^+)\ kg^{-1}$
Sand (light colour)	3-5
Sand (dark colour)	10-20
Loam	10-15
Silt loam	15-25
Clay and clay loam	20-50
Organic soil	50-100

Properties of clay minerals

Clay minerals	d space A^0	Shape	Y Index
Kaolinite	7.2	Plate like & Hexagonal	0-5
Halloysite	10.25	Elongated tubules	5-25
Illite	10	Irregular flake	-
Montmorillonite,	17-18	Irregular flakes	>25
Vermiculite, chlorite	14	Irregular flakes	-

Cation exchange capacity (CEC) of different clay minerals

Clay minerals	CEC cmol(p$^+$)kg^{-1}	Surface Area m^2 /Kg
Mica	0	-
Kaolinite	3-15	37-45
Muscovite	11	
Illite	30-40	120-170
Attapulgite	18-22	-
Nontronoite	57-64	-
Saponite	69-81	-
Monmorillonite	80-150	580-900
Vermiculite	150	780-900
Chlorite	-	130-180
Humus	200-300	

Available forms of nutrients:

Nitrogen	NO_3^- , NH_4^+
Phosphorus	$H_2PO_4^-$, HPO_4^{2-}
Sulphur	SO_4^{2-}, SO_3^{2-}
Boron	H_3BO_3, $H_2BO_3^-$, HBO_3^{2-}, BO_3^{3-}
Molybdenum	$Mo\,O_4^{2-}$

Mobility of Nutrient in soil

Mobile – NO_2/ NO^-_3, SO_4^{2-}, BO_3, Mn^{2+}, Cl^-

Less mobile: NH_4, K^+ Ca, Mg, Cu

Immobile : $H_2 PO_4^-$,$H PO_4$, Zn^{2+}, K^+, Mg^{2+}, Fe^{2+}

Mobility in pant

High mobile – N, P, K

Moderate – Zn

Less mobile – S, Fe, Cu, Mn, Mo & Cl

Immobile – Ca& B

C: N ratio	C: P ratio	Process
<20: 1	< 200:1 or 200:1	Mineralization
20 – 30:1	>200:1& <300: 1	No mineralization & Immobilization
>30: 1	300:1& above	Immobilization

Antagonistic effects in nutrients:

↑ Mg	↓ Fe, ↓ Ca, ↓ Cu
↑ Ca	↓ B , ↓ Fe
↑ K	↓ Mg , ↓ Mn
↑ P	↓ Zn , ↓ Fe

Excess of nutrient	Causes deficiency
Ca	P
Ca, Mg	K
K, NH₄	Mg
Fe, SO₄	Mo
P	Zn
N, P, K	Cu
NO₃	Fe
Zn, Al	Cu
N, K, Ca	B

Indicator plants for different nutrients

Elements	Indicator plants
N & Ca	Cauliflower & cabbage
P	Rape
K & Mg	Potato
Fe	Cauliflower
Na	Sugarbeet
Mn	Sugarbeet & oats
Boron	Sunflower

Elements	Function	Deficiency symptoms
Potassium	Regulates permeability of cells membrane Resistance to disease Necessary for sugar crops Activates enzymes Essential for photosynthesis	Dieback Leaf scorch Marginal chlorosis Rosette in beet , clelery carrot
Calcium	Stability of cell membrane Present as Calcium pectate in cell wall Neutralizes organic acids Prevent lodging of plants	Die back Tip hooking in cauliflower, beet Blossom end rot in tomato Black heart of cabbage Bitter pit of apple
Magnesium	Constituent of chlorophyll & chromosome Essential for Carbohydrate metabolism	Cotton reddening Bronzing of citrus Interveinal chlorosis (old leaves) Sand drown disease in tobacco Woolly flesh of tomato
Sulphur	Increase oil content in oil seed crops Responsible for Pungent odour of onion Promotes nodules formation Promotes root growth	Tea yellow disease Upward curling
Boron	Cell development Translocation of sugars Polysaccharide synthesis	Sterility & malformation Decrease the rate of water absorption Rosetting in Lucerne Snakehead in walnut Dieback& corking in apple Corking & pitting of tomato Hallow stem & bronzing of cauliflower Brown heart disease in table beets, turnip Heart rot of sugar beet Canker & internal black spot of garden pea Top sickness in tobacco Hard fruit of citrus Hen & chicken disease of grapes
Iron	Constituent of cytochrome Catalytic role in enzymatic activity	Interveinal chlorosis (young leaves) Mottle leaf in sugarcane Lime induced chlorosis

Contd..

Elements	Function	Deficiency symptoms
Manganese	Constituent of chlorophyll Oxidation reduction function Helps in iron movement	Gum packets in citrus Interveinal chlorosis Grey spec in oats, barley , wheat, rye, maize Pahala blight of sugarcane Marsh spot of peas Speckled yellow of sugarbeet
Zinc	Essential for chlorophyll synthesis Necessary for production of IAA(auxin) Necessary for production of cysteine Oxidation reduction function	Kaira disease of paddy White bud of maize Rosette of fruit trees Frenching of citrus Sickle leaf of cocoa Little leaf of apple
Copper	Necessary for synthesis of Vit A Catalyst in respiration	Multiple bud formation Gum pockets & dieback in citrus Fruit crack Exanthema of fruit trees Reclamation disease
Molybdenum	Constituent of nitrate reductase & nitrogenase Enhances Nitrogen fixation	Whip tail in cauliflower Scald of legumes Yellow spot of citrus Down ward cupping in raddish Affects nodulation

Problem soil and their properties

	Saline Soil	Alkali	Saline alkali	Degraded alkali
ESP (%)	<15	>15	>15	>15
pH	<8.5	8.5 –10.0	>8.5	8.5
TSS (%)	>0.1	<0.1	>0.1	<0.1
ECe	>4 dS / m	<4 dS/m	>4 dS/m	<4 dS/m

Component of soil colour

Hue	Denotes the dominant spectral colour (yellow, red, blue, brown)
Value	Denotes the lightness and darkness of colour
Chroma	Denotes purity of colour

Specific heat (cal. /gm)

Water	1
O.M	0.462
Clay	0.22
Air	24
Humus	0.4
Ice	0.5

Fertilizers Conversions:

P to P_2O_5	:	2.29	$ZnSO_4\ 7\ H_2O$ to Zn	:	0.227
P_2O_5 to P	:	0.436	MgO to Mg	:	0.603
K to K_2O	:	1.204	$MgSO_4\ 2H_2O$ to Mg	:	0.098
K_2O to K	:	0.830	N to NH_4	:	1.215
KCL to K	:	0.524	NO_3 to N	:	0.225
SO_2 to S	:	0.321	N to $(NH_4)_2SO_4$	:	4.716
Zn to $ZnSO_4\ 7\ H_2O$	:	4.39	Ca to CaO	:	1.40

Neutralizing value of liming materials

Liming materials		Neutralizing value or CCE (%)
Calcium Oxide (CaO)	:	179
Calcium hydroxide [Ca $(OH)_2$]	:	136
Dolomite	:	108.7
Calcite ($CaCO_3$)	:	100
Basic Slag ($CaSiO_3$)	:	86

Nutrient content of organic manures (%) (Approximation):

Organic Manures	N	P_2O_5	K_2O
Farm yard manure	0.5	0.3	0.5
Compost(urban)	1.0	0.5	0.5
Compost(Rural)	0.5	0.5	1.0
Cow dung	1.5	0.5	0.5
Poultry litter	3.0	2.5	1.5
Night soil	5.0	3.0	2.0
Castor cake	5.5	2.0	1.5
Groundnut cake	7.0	1.5	1.5
Cotton seed cake	6.0	3.0	2.0
Safflower cake	8.0	2.0	2.0
Bone meal	3.0	20.0	
Meat meal	10.5	2.5	0.5
Blood meal	10.0	1.5	1.0

SOIL MICROBES

Heat Requirements

- ✍ Psychrophiles   <10° C
- ✍ Mesophiles   20-40°C
- ✍ Thermophiles   >45° C

pH Requirements

- ✍ Fungi   4.5 - 6.5
- ✍ Bacteria   6.5 - 8.0

Symbiotic Nitrogen fixing bacteria: *Rhizobium*

Associative symbiosis: *Azospirillum*

Free living Nitrogen fixing bacteria: *Azotobacter, Beijerinkia, Clistridium*

Ammonifying bacteria: *Bacillus, Pseudomonas*

Nitrifying bacteria: *Nitrosomonas, Nitrobacter*

Denitrifying bacteria: *Pseudomonas, Bacillus, Paracoccus*

Cellulose decomposing bacteria: *Cytophaga, Sporocytophaga, Angiococcus, Polyangium, Clostridium*

Cellulose & hemi cellulose decomposing fungi: *Penicillium, Trichoderma*

Lignin decomposing fungi: *Basidiomycetes*

Protein decomposing fungi: *Fusarium, Aspergillus, Mucor*

Humus forming fungi: *Alternaria, Aspergillus*

Mycorrhizae: *Ecto Mycorrhizae* – Amanita, Boletus; *Endo Mycorrhizae* – Glomus, Endogene

Phosphate solubilising organisms: Bacteria – *Pseudomonas, Bacillus*; Fungi - *Penicillium, Aspergillus*

Gum producing Bacteria: *Azotobacter*, Beijerinkia, *Rhizobium*

Vermicompost

- ❖ *Eisenia foetida* - more efficient
- ❖ *Eudrillus engeniae*
- ❖ *Perionyx excavatus*

Culturing of earth warm is ***wormiri***

Constituents of biogas

CH_4	-	50-60 %
CO_2	-	30-40 %
H_2	-	5-10 %
N_2	-	4-6%

Transported Parent material

Agent	Deposited by (or) in	Name of deposit
Water	River Lake Ocean	Alluvium Lacustrine Marine
Wind	Wind	Loess –silt ,Aeolian – sand
Gravity	Gravity	Colluvium
Ice	Ice	Till , Moraine

Scientist and their contribution

Russel & Martin	Y index
Walkley & Black	Wet combustion method
Thompson & Later way	Ion Exchange
Wiegner	Reversibility & Equivalence of exchange Reaction
Adams & Holmes	Synthetic resin
Krishnamoorthy, Davis, Overstreet	Nearest Neighbour Model
Jenny	Oscillation volume
Kappan	Total acidity
Sokonov	Exchange acidity
Boussinghault	Legume can utilise the atmospheric Nitrogen
Beijerink	Isolate Rhizobium from root nodules
Kellner & Dyer	Fixation of potassium
E.J. Russel	Theory of bonding of clay- clay particle
Chepil	Dry sieving method
Tiulin & yoder	Wet sieving method
Kostiakov	Equation on infiltration
O'neal	Permeability class
Schofield	PF scale & ratio law
Ficks law	Gaseous exchange between soil & atmosphere

Contd..

Atterberg	Soil consistency limit
Arnon	Criteria of essentiality of nutrients
Nicholas	Functional or Metabolism nutrients
Sprengel	Calcium & magnesium
Salm – Horstmar	Sulphur
Warrington	Boron
Gris	Iron
Ma Hargue	Manganese
Sommer & Lipman	Zinc
Sommer	Copper
Arnon & stout	Molybdenum
Broyer	Chlorine
Bray & Kurtz	Phosphate estimation in Acid soil
Olsen	Phosphate estimation in slightly acid & Neutral soil
Das	Phosphate estimation in calcareous soil
Von Liebig	Law of Restitution & Law of minimum & law of limiting nutrient
Mitcherlich	Law of diminishing return & Pot culture Technique
Mehlich	*Aspergillus niger* method
Sackett & stewart	Soil plague method
Winogradsky	*Azotobacter* method
Schoonover	Gypsum requirement
Shoemaker	Lime requirement
Cate & Nelson	Critical soil test level approach
Dockuchaev	Factors of soil formation

Jenny	Classified the factors of soil formation as active & passive factors
Milne	Soil catena
Dockuchaev	Concept of zonality
Jenny	Expressed soil age in terms of time
Mohr & Baren	Five stages of soil formation
Buchanan	Coined the term Laterite
Vilenski	Climatic classification of soil
Virtanen	Part of Nitrogen fixed is excreted as free Amino acid
Dockuchaev & sibirtzev	Law of adaptability
Chepil & wood	Wind erosion equation

Contd..

Weinmaeir & smith	Universal soil equation
Puri	Salt index
Baldwin, kellog,thorp	Elobrated marbut's classification
Leather	First imperial agricultural chemist
Wadia, Krishnan & muderjee	Prepared soil map of India in 1935 based on geological formation
Schokalsky	Prepared soil map of India in 1932 showing 4 soil groups
Viswanath & ukil	Prepared soil map of India in 1943 showing 14 classes
Rey chaudhuri	Soils of India , their classification , occurrence, and properties
Murhty	Benchmark soils of India
Dokuchaev	Genetic system of classification
Marbut	Morphogenetic system of classification
Dachaufour	Coined the term Lessivage
Dalton	Atomic hypothesis
Neils bohr	Atomic structure for hydrogen
De broglie	Quantum mechanical model of atom
Doeberiner	Law of triads
Newlands	Law of octaves
Kossel & lewis	Octet rule
Heisenberg	Ortho & para hydrogen
Pierre & marrie curie	Discovered polonium & radium
Schmidt & curie	Thorium
Debierne	Actinium
Rutherford	Effect of electric & magnetic field on radioactive rays
Libby	Carbon dating
Thomson	Discovered electron
Rutherford	Discovered proton & spinthariscope
Chadwick	Coined the term neutron
Anderson	Positron
John Dalton	All matter was composed of atoms
Blackman	Law of limiting factor
Blackman	Electro chemical theory of nutrient uptake &theory of relativity
Nymith	Electro ultrafiltration theory
Ramamoorhty	Target yield concept
Ramamoorthy & subramanium	Equilibrium Phosphate potential

Contd..

Beckett	Phosphate equilibrium potential
Russel	E – value
Lorentz	Coined the term electron
Goldstein	Discovered positive rays
Bothe & Becker	Found neutron
Lawrence	Cyclotron
Henri Becquerel	Radioactivity
Curie & joliot	Artificial radioactivity
Pauli	Neutrino & anti neutrino
J.J. Thompson	Found e/m value
R.A. milikan	Found e value(electronic charge)
Rutherford	Alpha particle scattering experiment
Mosley	Determined atomic number or proton No.
Ottoman &stassmann	Nuclear fission
Cameroon	Soil is an inexhaustible source of nutrients
Baule	Percent yield concept & nutrient suffiency concept
Bray	Nutrient mobility concept & percent suffiency concept
Cate & Nelson	Critical soil test level approach
Nicholas	Ultra micronutrient concept
Marshall	CEC is next to photosynthesis & PH is physical phenomenon
Epstein	Carrier theory
S.V.Desai & S.C.Biswas	Biogas
La cassageue	Autoradiogaphy
Waksman	Lignin theory
Van Helment	Water essential for plant
Woodward	Water –transpiration
Lewis & randol	Ionic strength
Hevesy & hoffer	Isotopic dilution
Reinganum	Auto radioactivities
Jenny & overstreet	Root exchange
Devaux	Root CEC
Sorensen	pH concept
Silen	pE concept
Troug	Prescription approach & father of soil testing
Colwell	Integrated soil test approach
Beaufils	DRIS concept
Libby	Radio carbon dating

Contd..

Aslyng & latter schofield	Lime potential
Beckett	Q /I
Stevenson	Polyphenol theory of organic matter
Parker	Fertility or nutrient index
Lindsay &Norvell	DTPA extractable micro nutrients
Fried & Dean	A value
Larsen	Y or L value
Mitscherlich	Mathematical relation indicating relationship between plant growth response and growth factor
Chapman	Dignostic technique for soil and plants
Holmes	Total micronutrient
Lindsay & Norvell	Available nutrient
Wollvy	Father of soil physics
Dokuchaev	Normal and extra normal soil
Fallov	Residual and transported soil
Hilgard	Black alkali and white alkali soil
G.D. Smith	Seventh approximation
Kononova	Work on organic matter
Ramser	Rational formula for watershed
Andres	Liquefication of gas
J.B.Lawes	Super phosphate
Nernst	Third law of thermodynamics
Bragg	X ray diffraction
Murthy	Benchmark soil
Hole	Horizonaion
Kellog	Latosols
Buchanan	Laterite
Robinson	International pipette method
Atterberg	Plasticity limits

Water quality Parameters:

- Sodium Absorption Ratio(SAR) $= \dfrac{Na^+}{\sqrt{(Ca^{2+} + Mg^{2+})/2}}$

- Adj SAR=SAR $(1+ (8.4 - PH_c))$ (safe limit is <13)

- Salt index = (Total   Na – 24.5) – (Total Ca – Ca in $CaCO_3$) × 4.85

- (Safe limit is between -24.5 to 0)

- Residual Sodium carbonate RSC = $(CO_3^{\,2-} + HCO_3^{\,-}) - (Ca^{2+} + Mg^{2+})$
  Safe limit is < 1.25

- Chlorine index = $\dfrac{Cl + NO3}{CO32-+HCO3-+SO42-+Cl-+NO3}$
  (Safe limit is 4 to 12)

- Soluble Sodium percentage (SSP) $= \dfrac{Na^+}{Ca^{2+} + Mg^{2+} + Na^+} \times 100$
  (Safe limit is <60)

- Mg absorption Ratio $= \dfrac{Mg^{2+}}{Mg^{2+} + Ca^{2+}} \times 100$

- Potential salinity $= Cl + \frac{1}{2}\,SO_4$ (Safe limit is <5)

- Permeability index $= \dfrac{(Na^+ + HCO_3^{\,-})}{Ca^{2+} + Mg^{2+} + Na^+} \times 100$

- Total dissolved salt (TDS) = EC (dS/m)× 640 or EC (ms/m) × 10^{-2} × 640

- Osmotic pressure (OP) = 0.36× EC

- Millimhos / cm × 1 = Desisiemen / m

- Micromhos / cm × 0.1 = Millisiemen / m

Irrigation Water Quality

Particulars	EC (dSm^{-1})	pH	SSP	Cl$^-$ mel^{-1}	SAR	RSC mel^{-1}
Excellent	< 0.5	6.5-7.5	< 30	> 2.5	> 1	< 1.0
Good	0.5-1.5	7.5-8.0	30-60	2.5-5.0	1-2	1.0-1.25
Fair	1.5-3.0	8.0-8.5	60-75	5.0-7.5	2-4	1.25-2.0
Poor	3.0-6.0	8.5-9.0	75-80	7.5-10	4-8	2.0-2.5
Very poor	5.0-6.0	9.0-10	80-90	10-12.5	8-15	2.5-3.0
Unsuitable	> 6	> 10	> 90	> 12.5	> 15	> 3.0

Difference between Fulvic and Humic acid

Fulvic acid	Humic acid
Soluble in acid & alkali	Only alkali soluble
Yellow to brownish yellow	Dark brown to black
Mol wt few hundred to few thousand	10000 – 200000

Chemical nature of humic substances

Humic substance	Elemental composition (%)				
	C	H	N	S	O
Humic acid	54-59	3 –6	1 – 4	1 – 5	33 – 38
Fulvic acid	41 - 51	4 – 7	1 – 4	3 – 5	40 – 50

Functional group (c mol / kg)				
Humic substance	Total acids	Cool	Phenolic	E4/E6
Humic acid	500-900	150 – 600	200 – 600	3 –5
Fulvic acid	700- 1400	500 – 1100	110 – 600	6 – 9

Physical properties of Humus

❖ It posses low plasticity & cohesion

❖ It imparts the colour to soi

Colloidal properties

❖ They carry the negative charges

❖ It composed of C, H, O with trace amount of N, S, P & other element

❖ It adsorb Ca^{2+}, H^+, Mg^{2+}, K^+, Na^+

REMOTE SENSING

☞ Remote sensing is the sensing of objects or a phenomena without being in physical contact

☞ In remote sensing optical wave length from 0.3 to 15 μm are used

☞ Human eye responds to wave length between 0.4 to 0.7 μm and which is also photo synthetically active radiation range

☞ Near infra red wave length ranges between 0.7 to 1.0 μm and middle infra red ranges between 1.3 to 3.0 μm

☞ Velocity of light in air : 3×10^8 m/s

☞ Simple vegetation index = reflectance in infra red band / reflectance in red band

$$NDVI = \frac{Reflectance\ in\ IR\ band\ -\ Reflectance\ in\ Red\ Band}{Reflectance\ in\ IR\ band\ +\ Reflectance\ in\ Red\ Band}$$

☞ Vegetation index is used in identifying   the crop stress that can be used in crop management

☞ CARTOSTAT-2A launched in 2008 with a special resolution of less than 1 meter

☞ LISS stands for Linear Image Self Scanner, WiFS stands for Wide Field Scanner and PAN stands for Panchromatic Scanner

Characteristics of Indian Satellites

Parameter	IRS-1A / 1B	IRS-1C	IRS-1D
Height (Km)	904	817	817
Repetivity	22 days	24 days	25 days
Type	Sun-synchronous	Sun-synchronous	Sun-synchronous
Ground Resolution (m)	72.5 (LISS-I) 36.5 (LISS-II)	6.25 (PAN) 188 (WIFS)	5.2 (PAN) 169 (WIFS)

Application of Remote Sensing in Agriculture

Soil resource management

- ❖ Soil resource inventory
- ❖ Assessment of soil related constraints
- ❖ Development of management strategies

Crop health monitoring

- ❖ Assessment of drought
- ❖ Monitoring crop condition Fore warming short falls

Crop inventory

- ❖ Production forecasting
- ❖ Changes in crop pattern
- ❖ Cropping system analysis

Monitoring in season agricultural operation

- ❖ Progress of sowing total crop area
- ❖ Progress of harvest

Command area management

- ❖ Crop water requirement
- ❖ Water logging
- ❖ Crop violation
- ❖ Canal water seepage

Watershed management

- ❖ Characterization
- ❖ Development
- ❖ Prioritization
- ❖ Monitoring

Land use planning

- ❖ Assessment of land use and land covering
- ❖ Monitoring the land use changes
- ❖ Land evaluation or evaluation of potential land use option

Water availability

- ❖ Monitoring surface water bodies
- ❖ Ground water exploration

Land degradation

- ❖ Extent of waste land
- ❖ Problem soil (water logging and salt affected soil)

Damage assessment

- ❖ Damage due to cyclones and floods
- ❖ Large scale incident of pest and disease

Geographic Information System (GIS)

A geographic information system, commonly referred to as a GIS, is an integrated set of hardware and software tools used for the manipulation and management of digital spatial (geographic) and related attribute data for decision making planning and management of land use, natural resources, environment, transportation, urban facilities and others

Functions of GIS

Functions	Sub functions
Data acquisition and pre processing	Digitizing, Editing, Topology building, projection transformation, Format conversion, Attribute assignment.
Data base management and Retrieval	Data Archival, Hierarchical modeling, Network modeling, Relational modeling, Attribute query, Object –oriented attribute
Spatial measurement and analysis	Measurement operations , buffering, Overlay operations, Connectivity operations
Graphic output and visualization	Scale transformation, Generalization, Topographic map, Statistical map

Software used in Geographical Information System

- ☞ ArcGIS 9.2
- ☞ ArcINFO 8.0
- ☞ Idrisi 32.2.2
- ☞ Erdas Imagine 8.5
- ☞ Map INFO 7.8
- ☞ Carta LINX v2

Global Positioning System (GPS)

It is a satellite-based system that can be used to locate positions anywhere on the earth

- ❖ GPS used to identify the geographical co-ordinates associated with satellite imagery.
- ❖ GPS used in ground truthing of satellite images mainly co-ordinates of that region can be identified.

GPS system consists of three segments:

- ✍ Space segment
- ✍ Control system
- ✍ User segment

Books and Journals

Sl.no.	Journal	Publication
1.	Advances in soil science	Springer – verlag, berlin, new york.
2.	Australian journal of soil research	Commonwealth Scientific & Industrial Research Organisation (CSIRO) Australia.
3.	Communication in soil science and plant analysis	Marcel- Dekker ,Inc.
4.	Journal of Indian Society of Soil Science	ISSS, IARI, New Delhi, India.
5.	Plant & Soil (International Journal on Plant & Soil Relationship)	Kluwer academic Publishers, Netherlands.
6.	Soil Science (An Inter Disciplinary to Soil Research)	Lippincott Wiliams & Wilkins, Inc. USA.
7.	Canadian journal of soil science	The Agricultural Institute of Canada, Canada.
8.	Geoderma	Elsevier Science Publishers, Amsterdam, Netherlands.
9.	Pedologie	Belgium Society Of Soil Science, Krijgslaan. Belgium.
10.	Journal of soil science (British Society of Soil Science)	Blackwell Scientific Publications, London.

Contd..

Sl.no.	Journal	Publication
11.	Soil Science and Plant Nutrition	Japanese Society of Soil Science & Plant Nutrition. Japan.
12.	Soil & Tillage Research	Elsevier Publication, Amsterdam. Netherlands.
13.	Journal of plant nutrition and soil science	Willey-VCH.
14.	Soil Science Society of America Journa (SSSAJ)	Soil Science Society of America.
15.	Journal of Soil Biology & Ecology	Indian Society of Soil Biology & Ecology, UAS, Bangalore. India.
16.	Journal of Soil & Water Conservation	Soil & Water Conservation Society, Ankeny.
17.	Indian journal of soil conservation	Association of soil & water conservation.
18.	Soil Biology And Biochemistry	Elsevier Publication, Amsterdam. Netherlands.

Conversion factors

Sl. No	Parameter		Equation/ conversion factors
1	eq.wt.	=	$\dfrac{\text{Atomic mass}}{\text{Electrovalence}}$
2	Molarity	=	$\dfrac{\left(\text{wt.}/\text{L}\right)}{\text{mol.wt}}$
3	Normality	=	$\dfrac{\left(\text{wt.}/\text{L}\right)}{\text{Equivalent wt.}}$
4	Normality of acid	=	$\dfrac{\text{SG}\times\text{assay}\times1000}{\left(\text{eq.wt}\times100\right)}$
5	Molality	=	$\dfrac{\left(\text{wt.}/1000\text{g}\right)}{\text{mol.wt}}$
6	m.eq.	=	$\dfrac{\text{Equivalent weight}}{1000}$
7	m. eq.	=	TV (mL) × N
8	m.eq × eq.wt	=	Mg

Contd..

Sl. No	Parameter	=	Equation/ conversion factors
9	m.eg × meq.wt	=	G
10	(m.eq / L) × eq.wt	=	ppm
11	m.eq. / L		$\dfrac{ppm}{eq.wt}$
12	ppm	=	mg / L
13	ppm	=	% × 10000
14	ppm x 2	=	lb/ acre = kg/ ha
15	% = g / 100 g	=	$\dfrac{ppm}{10000}$
16	%	=	(m.eq. / 100 g) × meq. wt
17	ppm	=	$0.64 \times EC \times 10^6$
18	% moisture in saturated paste	=	$\dfrac{(2.65 \times volume \times mass) \times 100}{2.65 \times (mass - volume)}$
19	ECe (soil saturated paste) (msp)	=	$\dfrac{ECs\ (1:2) \times 200}{\% \text{ moisture in saturated paste}}$
20	% salt in soil	=	$0.064 \times EC \times 1000 \times \% \text{ msp}$
21	mhos / cm × 10^3	=	milli mhos / cm
22	mhos / cm × 10^6	=	micro mhos / cm
23	1 acre	=	4840 sq. yds. = 4047 sq. m = 43560 sq.ft. = 0.404686 ha
24	1 ha	=	2.471 acre or 10000 m^2
25	1 lb	=	0.4536 kg
26	1 kg	=	2.2046 lbs
27	1 acre furrow slice (6"depth)	=	2 million pounds (m. lbs)
28	1 hectare furrow slice (6"depth)	=	2 million kg
29	1 micron(µ)	=	10^{-4} cm.

Contd..

Sl. No	Parameter		Equation/ conversion factors
30	1 milli micron (mμ)	=	10^{-7} cm.
31	1 A^0 (Angstrom)	=	10^{-8} cm.
32	% element	=	$\dfrac{\% \text{ compd.} \times \text{at. wt. of element} \times \text{no. of elements}}{\text{mol. wt. of compd.}}$
33	% P = % P_2O_5 x 0.44	=	$\dfrac{\% \ P_2O_5 \times 62}{142}$
34	% compound	=	$\dfrac{\% \text{ element} \times \text{mol. wt of compd.}}{\text{at. wt. element} \times \text{no. of elements}}$
35	% P_2O_5 = % P x 2.29	=	$\dfrac{\%P \times 142}{62}$
36	Desired Compound equivalence (tons ha^{-1})	=	$\dfrac{\text{Ref. compd.} \times \text{mol. wt of desired compd.}}{\text{mol. wt of ref. compd.}}$
37	CCE (Calcium carbonate equivalent)	=	$\dfrac{1 \times 56 \ (CaO)}{100 \ (CaCO_3)} = 0.56$
36	1 cmol/ kg	=	1 meq/ 100 g
37	1 mmhos/ cm at 25°C	=	1 dS/ m at 25°C
38	1 meq of H^+ / 100 g	=	1.00 tons of Lime ($CaCO_3$)/ ha
39	1 meq of Na^+ / 100 g	=	1.72 tons of Gypsum ($CaSO_4.2H_2O$)/ ha
40	1 mL of 1 N $K_2Cr_2O_7$	=	0.003 g Carbon

Strength of aqueous solutions of common acids and bases

Reagent	Weight (%)	Sp. Gravity (g/CC)	Normality	Volume required for 1 L of 1N soln.
HCl	35.0	1.18	11.3	89
HNO_3	70.0	1.42	16.0	63
H_2SO_4	96.0	1.84	36.0	28
H_3PO_4	85.0	1.69	41.0	23
CH_3COOH	99.5	1.05	17.4	58
NH_4OH	27.0	0.90	14.3	71

Indicators used for acid base titrations

Sl. No.	Titration	Indicators
1	Strong acid v/s strong base	phenolphthalein, methyl red or methyl orange
2	Strong acid v/s weak base	Methyl orange or methyl red
3	Weak acid v/sstrong base	Phenolphthalein

Preparation of some indicators

1. **Phenolphthalein:** Dissolve 0.5 g of phenolphthalein in 50 mL of alcohol and add 50 mL of distilled water with constant stirring.

2. **Methyl orange:** Dissolve 0.5 g of free acid or of sodium salt in 100 mL of distilled water and filter the solution.

3. **Methyl red:** Dissolve 0.1 g of methyl red in 100 mL of hot water or dissolve in 60 mL of alcohol and dilute with 40 mL of distilled water.

4. **Bromo thymol blue:** Grind 1 g of bromothymol blue and dissolve in 100 mL of alcohol.

5. **Mixed indicator:** Dissolve 0.1 g of bromo cresol green and 0.07 g of methyl red in 100 mL of alcohol.

6. **Diphenylamine:** Dissolve 0.5 g of diphenyl amine in a mixture of 20 mL water and 100 mL conc. sulphuric acid.

7. **Potassium chromate:** Dissolve 5 g of AR potassium chromate in 100 mL distilled water.

8. **Mureoxide:** Mix 0.2 g of ammonium purpurate with 40 g of potassium sulphate and grind to a fine powder using glass/ agate pestle and mortar.

9. **EBT (Erichrome Black-T):** Dissolve 0.2 g of EBT in 50 mL of methanol.

10. **Bromo cresol Green:** Dissolve 1.0 g of bromo cresol green in 100 mL alcohol.

11. **Starch:** Take 1 g of starch in hot water and make a paste and then pour it in 100 mL of boiling water, mix for 2 minutes, cool it and then add 3.0 g of KI. Add a pinch of mercuric iodide or boric acid to avoid microbial decomposition.

12. **Ferroin:** Dissolve 695 mg of ferrous sulphate $(FeSO_4.7H_2O)$ in 100 mL distilled water. Add 1.485 g of 1, 10- Ortho phenanthroline monohydrate and mix thoroughly until dissolution occurs.

13. **Patton and Reeders reagent:** Mixture of 0.5 g of the pure Patton-Reeder indicator [2-hydroxy-1-(2-hydroxy-4-sulfo-1-naphthylazo)-3-naphthoic acid] and 50 g of sodium sulfate ground together to a fine powder.

Preparation of standard solution

Sl. No.	Reagent	To prepare 1 L of approx. 0.1 N soln.	To prepare 1 L of 0.1 N soln. used for Standardization	Indicators and colour change
1	Sulphuric acid	Dilute 3.0 mL of conc. H_2SO_4 to 1 L with distilled water	0.1 N N Na_2CO_3: Dissolve 5.29 g of AR grade Na_2CO_3 in 1 L or 0.1 N borax: dissolve 19.07 g of pure $Na_2B_4O_7.10 H_2O$ in 1 L	Methyl orange, yellow to orange
2	Nitric acid	Dilute 6.3 mL of Conc. HNO_3 to 1 L with distilled water	0.1 N N Na_2CO_3 or 0.1 N borax	Methyl orange, yellow to orange
3	Hydrochloric acid	Dilute 10 mL of conc. HCl to 1 L distilled water	0.1 N potassium hydrogen phthalate: Dissolve 20.4 g of potassium hydrogen phthalate in 1 L	Methyl orange, yellow to orange
4	Sodium hydroxide	Dissolve 4 g of NaOH in distilled water and make up to 1 L	0.1 N succinic acid: Dissolve 5.9 g of AR grade succinic acid in 1 L	Phenolphthalein, Colourless to light pink
5	Potassium hydroxide	Dissolve 5.6 g of KOH in distilled water and make up to 1 L	0.1 N succinic acid: Dissolve 5.9 g of AR grade succinic acid in 1 L	Phenolphthalein, Colourless to light pink
6	Silver nitrate	Dissolve 16.99 g of crystallized $AgNO_3$ and make up to 1 L of distilled water	0.1 N NaCl: Dissolve 5.846 g of AR grade NaCl in distilled water and make up to 1 L	Potassium dichromate, yellow to brick red ppt

Contd..

Sl. No.	Reagent	To prepare 1 L of approx. 0.1 N soln.	To prepare 1 L of 0.1 N soln. used for Standardization	Indicators and colour change
7	Potassium dichromate	Dissolve 4.904 g of $K_2Cr_2O_7$ in distilled water and make up to 1 L	0.1 N ferrous solution: 0.25 g of pure iron, 20 g of $NaHCO_3$ and 20 mL of 5 N H_2SO_4 and make up to 1 L	Diphenyl amine, brown to green
8	Potassium permanganate	Dissolve 4.904 g of $KMnO_4$ and make up to 1 L	0.1 N oxalic acid: Dissolve in 6.29 g of oxalic acid in distilled water and make up to 1 L or 0.1 N sodium oxalate: Dissolve 6.69 g of sodium oxalate In distilled water and make up to 1 L (titration carried out to $70\text{-}80^0C$)	Self indicator, Colourless to pink
9	Iodine	Dissolve 12.7 g of iodine in saturated solution of potassium iodide and transfer the soluble iodine to volumetric flask and make up to 1 L	0.1 N sodium thiosulphate: Dissolve 24.819 g sodium thiosulphate in distilled water and make up to 1 L	Starch, Colour changes from blue to milky white

Plant parts recommended for sampling

	Crop	Part to be sampled with stage/ age
I	**Cereals, pulses, oilseeds, fibre and commercial crops**	
1	Wheat	Flag-leaf, before head emergence
2	Rice	3rd leaf from apex, at tillering
3	Maize	Ear-leaf, before tasseling
4	Barley	Flag-leaf at head emergence
5	Oat	Flag-lea, before inflorescence emergence
6	Pulses	Recently matured leaf, at bloom initiation
7	Groundnut	Recently manufactured leaflets, at maximum tillering
8	Sunflower	Youngest mature leaf blade, at initiation of flowering
9	Mustard	Recently matured leaf, at bloom initiation
10	Soybean	3rd leaf from top, after 2 months of planting
11	Cotton	Petiole, 4th leaf from the apex, at initiation of flowering

Contd..

	Crop	Part to be sampled with stage/ age
12	Jute	Recently matured leaf, at 60 days age
13	Sugarcane	3rd leaf from top, after 3-5 months of planting
14	Sugar beat	Petiole of youngest mature leaf, at 50-80 days age
15	Tea	3rd leaf from tip of young shoots
16	Coffee	3rd and 4rh pair of leaf from apex of lateral shoots, at bloom

II		Vegetable crops
1	Potato	Most recent, fully developed leaf (half-grown)
2	Tomato	Leaves adjacent to inflorescence (mid-bloom)
3	Onion	Top non-white portion(1/3 to ½ grown)
4	Brinjal	Blade of most recent, fully developed leaves
5	Beans	Uppermost, fully developed leaves
6	Cauliflower	Most recent, fully matured leaf, at heading
7	Cabbage	Wrapper leaf at 2-3 months age
8	Pea	Leaflets from most recent, fully developed leaves, at first bloom
9	Carrot	Most recent, fully matured leaf, at mid-growth
10	Radish	Most recent, fully developed leaf
11	Turnip, Sugar beet, Spinach	Most recent, fully developed leaf, at 30-50 days age 5th leaf from tip (omit unfurled) at the stage of bud starting to small fruits
12	Cucumber	Most recent, fully developed leaf

III	Ornamental Plants	
	Bougainvillea, Jasmine, Croton, Fern, Ficus, Geranium, Gerbera, Gladiolus, Lilly, Orchid, Hibiscus, Rose	Most recent, fully developed leaves at die stage of flower bud of pea size.

IV	Fruit crops	
1	Almond	3rd leaf from top, at the beginning of bloom
2	Apple, Pear	Leaves from middle of terminal shoot growth, 8-12 weeks after full bloom, 2-4 weeks after formation of terminal buds in bearing trees
3	Blackberry	Latest matured leaf from non-tipped canes, 4-6 weeks after peak bloom

Contd..

4	Cherry	Fully expanded leaves, mid shoot current growth in July- August
5	Peach	Mid-shoot leaves, fruiting or non-fruiting spurs, mid- summer leaves, fruiting
6	Strawberry	Youngest, fully expanded matured leaf with out petiole, at peak or harvest period
7	Plum	Leaves from middle of current season's extension growdi, in January-February
8	Banana	Petiole of 3rd open leaf from apex, 4 months after planting
9	Cashew	4th leaf from tip of matured branches, at beginning of flowering
10	Grapes	5th petiole from base at bud differentiation for yield, and petiole opposite to bloom for quality
11	Citrus fruits	3-5 months old leaves from new flush, 1st leaf of the shoot, in June
12	Guava	3rd pair of recently matured leaves, at bloom (August or December)
13	Mango	Leaf with petiole (4-7 months old) from middle of shoot
14	Papaya	6th petiole from apex, six months after planting
15	Pineapple	Middle one-third portion of white basal portion of 4th leaf from apex, at 4-6 months age
16	Pomegranate	8th leaf from apex at bud differentiation, in April and August
17	Falsa	4th leaf from apex, one month after pruning
18	Ber	6th leaf from apex from secondary or tertiary shoot, 2 months after pruning

Sources: Kensworthy (1964), Reuter and Robinson (1986), Jones et al., (1991), Bhargava and Raghupathi (1993)

EXCERSISE - 1

1. The components of fine earth are
   (a) Stones, sand, clay
   (b) Sand, silt ,clay
   (c) Gravel, sand, silt
   (d) Stones, gravel, sand

2. The zinc bearing mineral is
   (a) Orthoclase
   (b) Pyrite
   (c) Siderite
   (d) Sphalerite

3. The chemical weathering processes reduction is occurred in
   (a) Flooded condition
   (b) Dry condition
   (c) moist condition
   (d) High O_2 supply condition

4. The dominant soil order in india
   (a) Vertisol
   (b) Inceptisol
   (c) Entisol
   (d) Oxisol

5. In montmorillonite , negative charges arises mostly by
   (a) Decrease in pH
   (b) Isomorphous substitution
   (c) Broken edges
   (d) Exposed OH groups

6. The infiltration rate in the soils is as follows
   (a) Loam > clay > sand
   (b) sand > Loam > Clay
   (c) Clay > loam > sand
   (d) sand < Loam < Clay

7. Tensiometer works upto a tension of
   (a) 1.0 bar
   (b) 1.5 bar
   (c) 0.85 bar
   (d) 1.25 bar

8. Exchangeable cation responsible for deflocuulation of clay is
   (a) Ca
   (b) Na
   (c) Mg
   (d) Zn

9. Laterite soils belongs to the order
   (a) Vertisol
   (b) Inceptisol
   (c) Entisol
   (d) Oxisol

10. Which one of the following is having highest molecular weight
   (a) Fulvic acid
   (b) Humin
   (c) Humic acid
   (d) Hymatomelanic acid

11. Plasticity is highest for
   (a) Coarse Sand
   (b) Fine Sand
   (c) Silt
   (d) Clay

12. The mineral which forms the characteristic component of rocks are known as
    (a) Secondary minerals
    (b) Accessory minerals
    (c) Primary minerals
    (d) Essential minerals

13. The second most dominant element in the earth crust is
    (a) Oxygen   (b) Aluminium   (c) silica   (d) Iron

14. The water content at suction of -15 bars is termed as
    (a) Permanent wilting point
    (b) Hygroscopic Coefficient
    (c) Field Capacity
    (d) Saturation

15. The master horizon A indicates
    (a) organic horizon
    (b) Parent material
    (c) Mineral horizon
    (d) Bed rock

KEY TO EXERSISE - 1

1. b	2. d	3. a	4. b	5. b	6. b
7. c	8. b	9. d	10. b	11. d	12. d
13. c	14. a	15. c			

EXCERSISE - 2

1. The nutrient which is lost maximum by leaching is
    (a) Nitrogen   (b) Phosphorous   (c) Potassium   (d) Magnesium

2. Reduction in crop yields without appearance of any definite symptom is known as
    (a) sufficiency
    (b) Deficiency
    (c) Hidden hunger
    (d) Luxury consumption

3. The secondary nutrient which is relatively mobile in soils and immobile in plants is
    (a) Ca   (b) K   (c) Mg   (d) N

4. If a soil test value of a nutrient is low, then response for that nutrient is
    (a) low   (b) medium   (c) high   (d) none

5. Rapid tissue test for determination of phophates is
    (a) Olsen reagent
    (b) Sodium cobalt nitrate
    (c) Molybdate+ stannous oxalate
    (d) Diphenylamine

6. Among the following crops, one in which concentration of calcium is high

   (a) rice   (b) jowar   (c) bajra   (d) ground nut

7. Biological N fixation is a

   (a) Oxidation   (b) Reduction

   (c) Conjugation   (d) Hydration

8. The element essential for plant nutrient

   (a) Na   (b) Ni   (c) Cl   (d) Si

9. Chemical reagent used as an extractant for available cationic micronutrients is

   (a) DTPA + Ca Cl_2 (PH 7.3)   (b) Hot water

   (c) Netral normal ammonium acetate   (d) Olsens reagent

10. The Mulders *Aspergillus niger* test for evaluating the nutrient is

    (a) N   (b) K_2O (c) Cu and Mg (d) P_2O_5

11. Symptoms of Mn deficiency in peas is known as

    (a) pahalablight   (b) marsh spot

    (c) speckled yellow   (d) grey speck

12. The main structural constituents of plants are

    (a) C   (b) H   (c) O   (d) All the above

13. The Predominant micronutrient deficiency in soils of A.P is ————

    (a) Mn   (b) S   (c) Zn   (d) Cu

14. The secondary nutrient that can be absorbed by the soil directly from atmosphere is ————

    (a) Ca   (b) Mg   (c) S   (d) N

15. The ammendment most suitable for reclamation of calcareous alkali soils is ————

    (a) Pyrites   (b) Gypsum   (c) Lime   (d) None of these

KEY TO EXERSISE - 2

1. a	2. c	3. a	4. c	5. c	6. d
7. b	8. c	9. a	10. c	11. b	12. d
13. c	14. c	15. a			

EXCERSISE - 3

1. The term used to indicate minimum guaranteed quantity of major nutrients in multi nutrient fertilizer is
   - (a) Fertilizer Ratio
   - (b) Fertilizer Grade
   - (c) Fertilizer regulation
   - (d) Fertilizer mixture

2. Murate of potash is not recommended for one of these crops
   - (a) Rice
   - (b) Jowar
   - (c) Tobacco
   - (d) Maize

3. Phosphorus use efficiency can be increased by
   - (a) Band placement
   - (b) Split application
   - (c) Coated urea
   - (d) All of the above

4. An example of insect attractants is
   - (a) Methyl eugenol
   - (b) Zno
   - (c) Methyl Bromide
   - (d) Phygon

5. Mode of action of carbendazim on
   - (a) Inhibiting the DNA Synthesis
   - (b) Inhibiting the Photosynthesis
   - (c) Inhibiting the Respiratory enzymes
   - (d) Inhibiting the hydrolytic enzymes

6. Application of organic manure improves the
   - (a) Soil structure
   - (b) Soil aeration
   - (c) Water holding capacity
   - (d) all the above

7. Calcium deficiency in soils is common in
   - (a) low rain fall areas
   - (b) Arid region
   - (c) salt affected soils
   - (d) high rain fall areas

8. The complex fertilizer that is marketed in liquid form
   - (a) APS
   - (b) APP
   - (c) DAP
   - (d) UAP

9. Oxime functional group is represented as
   - (a) SH
   - (b) $C = O$
   - (c) $= C = N - O$
   - (d) $- C = N.$

10. An imoportant selective herbicide for Maize crop is
    - (a) Glyposate
    - (b) Paraquot
    - (c) Atrazine
    - (d) Diuron

11. Optical isomerism is exhibited by those compounds having
    - (a) Isometric carbon
    - (b) Symmetric carbon
    - (c) Asymmetric carbon
    - (d) None of these

12. Among the following fertilizers, the highest Nitrogen content fertilizer is

(a) Urea

(b) Ammonium sulfate

(c) Aqua Ammonia

(d) Gaseous Ammonia

13. Which of the following fungicide can be used as seed dresser and seed disinfection is

(a) Ethrimol

(b) Topsin –M

(c) Captan

(d) Carboxin

14. Diazinon is insecticide, which belongs to

(a) Thiophosporic acid derivatives

(b) Dithiophosporic acid derivatives

(c) Phosporic acid derivatives

(d) Pyrophosporic acid derivatives

15. DDVP is a

(a) Systemic poision

(b) Contact poision

(c) Stomach poision

(d) Stomach & Contact poision

KEY TO EXERSISE - 3

1. b	2. c	3. a	4. a	5. a	6. d
7. d	8. b	9. c	10. c	11. c	12. d
13. c	14. a	15. d			

EXCERSISE - 4

1. Who proved that air and water were the primary sources of C, H and O in plant tissue?

(a) Justus Von Leibig

(b) J.B. Boussingault

(c) Warrington

(d) F.H. king

2. The zinc bearing mineral is

(a) Orthoclase

(b) Pyrite

(c) Siderite

(d) Sphalerite

3. The chemical weathering processes reduction is occurred in

(a) Flooded condition

(b) Dry condition

(c) Moist condition

(d) High O_2 supply condition

4. Secondary silicate minerals are prominent in

(a) Silt fractions

(b) Sand fractions

(c) Sand and silt fractions

(d) Clay fractions

5. In montmorillonite , negative charges arises mostly by
   - (a) Decrease in pH
   - (b) Isomorphous substitution
   - (c) Broken edges
   - (d) Exposed OH groups

6. The infiltration rate in the soils as follows
   - (a) Loam > Clay > Sand
   - (b) Sand > Loam > Clay
   - (c) Clay > Loam > Sand
   - (d) Sand < Loam < Clay

7. Tensiometer works upto a tension of
   - (a) 1.0 bar
   - (b) 1.5 bar
   - (c) 0.85 bar
   - (d) 1.25 bar

8. Exchangeable cation responsible for deflocculation of clay is
   - (a) Ca
   - (b) Na
   - (c) Mg
   - (d) Zn

9. Laterite soils belongs to the order
   - (a) Vertisol
   - (b) Inceptisol
   - (c) Entisol
   - (d) Oxisol

10. Which one of the following is having highest molecular weight
    - (a) Fulvic acid
    - (b) Humin
    - (c) Humic acid
    - (d) Hymatomelanic acid

11. Plasticity is highest for
    - (a) Coarse Sand
    - (b) Fine Sand
    - (c) Silt
    - (d) Clay

12. The mineral which forms the characteristic component of rocks are known as
    - (a) Secondary minerals
    - (b) Accessory minerals
    - (c) Primary minerals
    - (d) Essential minerals

13. The second most dominant element in the earth crust is
    - (a) Oxygen
    - (b) Aluminium
    - (c) Silica
    - (d) Iron

14. The water content at suction of -15 bars is termed as
    - (a) Permanent wilting point
    - (b) Hygroscopic Coefficient
    - (c) Field Capacity
    - (d) Saturation

15. If bulk density and particle density 1.40 Mg m-3 and 2.80 Mg m-3 respectively the percentage pore space will be
    - (a) 40
    - (b) 50
    - (c) 60
    - (d) none of these

16. The dominant soil order in India
    - (a) Vertisol
    - (b) Inceptisol
    - (c) Entisol
    - (d) Oxisol

17. The master horizon A indicates
    (a) Organic horizon
    (b) Parent material
    (c) Mineral horizon
    (d) Bed rock

18. Addition of organic matter
    (a) decrease the bulk density
    (b) increase the porosity
    (c) both
    (d) None of these

19. The components of fine earth are
    (a) Stones, sand, clay
    (b) Sand, silt ,clay
    (c) Gravel, sand, silt
    (d) Stones, gravel, sand

20. The main effect of soil compaction is
    (a) Increasing bulk density
    (b) decreasing micro pores
    (c) increasing particle density
    (d) both b & c

21. The fraction of humus that is soluble in both acid and alkali is
    (a) Humin   (b) Humic acid   (c) Fulvic acid (d) both a and b

22. The degree of saturation is >80%, it indicates
    (a) Highly fertile soil
    (b) Medium fertile
    (c) Low fertile
    (d) None of these

23. The dominant ions under acid conditions are
    (a) Ca & Mg   (b) K &Na   (c) Al & H   (d) H&OH

24. Kaolinite mineral cannot expand because of the presence of the following
    (a) H bonding
    (b) K bonding
    (c) Brucite layer
    (d) Divalent ions

25. Clay mineral Illite belongs to
    (a) 1: 1   (b) 2: 1   (c) 2:1:1   (d) 2:2

KEY TO EXERSISE - 4

1. b	2. d	3. a	4. d	5. b	6. b
7. c	8. b	9. d	10. b	11. d	12. c
13. c	14. a	15. b	16. b	17. a	18. c
19. b	20. a	21. c	22. a	23. c	24. a
25. b					

EXCERSISE - 5

1. The nutrient which is lost maximum by leaching is
   (a) Nitrogen   (b) Phosphorous   (c) Potassium   (d) Magnesium

2. The extractant used for determining the available phosporous for acid soil is
   (a) Neutral Normal ammonium acetatate
   (b) Olsen reagent
   (c) Bray & Kurtz
   (d) Potassium permanganate

3. Reduction in crop yields without appearance of any definite symptom is known as
   (a) sufficiency   (b) Deficiency
   (c) Hidden hunger   (d) Luxury consumption

4. The secondary nutrient which is relatively mobile in soils and immobile in plants is
   (a) Ca   (b) K   (c) Mg   (d) N

5. The per cent of nitrogen in soils as inorganic source is
   (a) 98   (b) 2   (c) 50   (d) 75

6. If a soil test value of a nutrient is low, then response for that nutrient is
   (a) low   (b) medium   (c) high   (d) none

7. Zinc deficiency of paddy is known as
   (a) Khaira disease   (b) Akiochi
   (c) Grass tetany   (d) Top sickness

8. Rapid tissue test for determination of Nitrates is
   (a) Olsen reagent   (b) Sodium cobalt nitrate
   (c) Molybdate + stannous oxalate   (d) Diphenylamine

9. Among the following crops, one in which concentration of calcium is high
   (a) Rice   (b) Jowar   (c) Bajra   (d) Ground nut

10. Biological N fixation is a
    (a) Oxidation   (b) Reduction
    (c) Conjugation   (d) Hydration

11. The element essential for plant nutrient
    (a) Na   (b) Ni   (c) Cl   (d) Si

12. Chemical reagent used as an extractant for available cationic micronutrients is

   (a) DTPA + Ca Cl$_2$ (PH 7.3)   (b) Hot water

   (c) Netral normal ammonium acetate   (d) Olsens reagent

13. The Mulders *Aspergillus niger* test for evaluating the nutrient is

   (a) N   (b) K$_2$O   (c) Cu and Mg   (d) P$_2$O$_5$

14. Symptoms of Mn deficiency in peas is known as

   (a) pahalablight   (b) marsh spot

   (c) speckled yellow   (d) grey speck

15. The main structural constituents of plants are

   (a) C   (b) H   (c) O   (d) All the above

16. The Predominant micronutrient deficiency in soils of (A)P is————

   (a) Mn   (b) S   (c) Zn   (d) Cu

17. The secondary nutrient that can be absorbed by the soil directly from atmosphere is ————

   (a) Ca   (b) Mg   (c) S   (d) N

18. The percent recovery of phosphorous for typical black soils is obtained by the equation

   (a) 100 - clay %   (b) 100 – 2 clay %

   (c) 100 – 3 clay %   (d) 100 – 1/2clay %

19. The amendment most suitable for reclamation of calcareous alkali soils is

   (a) Pyrites   (b) Gypsum   (c) Lime   (d) None of these

20. The crop that high salt tolerant crop is

   (a) sugar beet   (b) tobacco   (c) citrus   (d) cauliflower

21. Due to over liming these soils can be formed

   (a) Acid soils   (b) Calcareous soils

   (c) Sodic soils   (d) Alkali soils

22. E.C of irrigation water is 2 dSm^{-1} and SAR is 5, this water belongs to

   (a) C$_1$S$_2$ (b) C$_2$S$_3$ (c) C$_3$S$_1$ (d) C$_4$S$_3$

23. Due to high EC of saline soil, the effect on crop is

   (a) Root growth   (b) Stem growth

   (c) Germination of seeds   (d) opening of flowers

24. Under following condition Nitrogen not available to the plants mainly due to
    - (a) Mineralization
    - (b) Denitrification
    - (c) Immobolizaion
    - (d) both b & c

KEY TO EXERSISE - 5

1. a	2. b	3. c	4. a	5. b	6. c
7. a	8. d	9. d	10. b	11. c	12. a
13. c	14. b	15. d	16. c	17. c	18. a
19. a	20. a	21. b	22. c	23. c	24. d

EXCERSISE - 6

1. The starter used in vermicompost is
   - (a) Night soil
   - (b) Dung
   - (c) Sullage
   - (d) Urea

2. When any organic manure is fermented under presence of oxygen ,then the gas produced is
   - (a) CH_4
   - (b) NH_3
   - (c) O_2
   - (d) CO_2

3. The term used to indicate minimum guaranteed quantity of major nutrients in multi nutrient fertilizer is
   - (a) Fertilizer Ratio
   - (b) Fertilizer Grade
   - (c) Fertilizer regulation
   - (d) Fertilizer mixture

4. The percent Nitrogen content in DAP is
   - (a) 46
   - (b) 18
   - (c) 21
   - (d) 46

5. Continuous application of Ammonium sulphate to soil result in
   - (a) Development of acidity in soil
   - (b) Development of salinity in soil
   - (c) Development of alkali soils
   - (d) None of these

6. Murate of potash is not recommended for one of these crops
   - (a) Rice
   - (b) Jowar
   - (c) Tobacco
   - (d) Maize

7. Phosphorus use efficiency can be increased by
   - (a) Band placement
   - (b) Split application
   - (c) Coated urea
   - (d) All of the above

8. An example of insect attractants is

   (a) Methyl eugenol (b) Zno

   (c) Methyl Bromide (d) Phygon

9. Mode of action of carbendazim on

   (a) Inhibiting the DNA Synthesis

   (b) Inhibiting the Photosynthesis

   (c) Inhibiting the Respiratory enzymes

   (d) Inhibiting the hydrolytic enzymes

10. Application of organic manure improves the

    (a) Soil structure (b) Soil aeration

    (c) Water holding capacity (d) all the above

11. Calcium deficiency in soils is common in

    (a) low rain fall areas (b) Arid region

    (c) salt affected soils (d) high rain fall areas

12. The complex fertilizer that is marketed in liquid form

    (a) APS (b) APP (c) DAP (d) UAP

13. Oxime functional group is represented as

    (a) SH (b) $C = O$

    (c) $= C = N - O$ (d) $- C = N.$

14. An imoportant selective herbicide for Maize crop is

    (a) Glyposate (b) Paraquot (c) Atrazine (d) Diuron

15. Optical isomerism is exhibited by those compounds having

    (a) Isometric carbon (b) Symmetric carbon

    (c) Asymmetric carbon (d) None of these

16. Among the following fertilizers, the highest Nitrogen content fertilizer is

    (a) Urea (b) Ammonium sulfate

    (c) Aqua Ammonia (d) Gaseous Ammonia

17. The main defect of natural pyrethrum that limits its use in Agricultural fields is its

    (a) High mammalian toxicity

    (b) Rapid photo degradation

    (c) Low insecticidal activity

    (d) High persistence

18. Which of the following fungicide can be used as seed dresser and seed disinfection is

    (a) Ethrimol   (b) Topsin –M   (c) Captan   (d) Carboxin

19. Diazinon is insecticide, which belongs to

    (a) Thiophosporic acid derivatives

    (b) Dithiophosporic acid derivatives

    (c) Phosporic acid derivatives

    (d) Pyrophosporic acid derivatives

20. DDVP is a

    (a) Systemic poision   (b) Contact poision

    (c) Stomach poision   (d) Stomach & Contact poision

21. Methyl benzimidazole carbomate is the Chemical name of

    (a) Ethrimol   (b) Carbendazim

    (c) Capton   (d) Carboxin

22. Butachlor is extensively used in

    (a) Rice   (b) Maize   (c) Jowar   (d) Cotton

23. Non selective pre emergence herbicide which also act as a desiccant and defoliant in cotton is

    (a) 2,4-D   (b) Paraquat   (c) Diuron   (d) Atrazine

24. The phosphorus fertilizer suitable for long duration crops in acid soil is

    (a) SSP   (b) DAP

    (c) Rock phosphate   (d) TSP

25. The Cost of 1 quintal SSP is 784 Rs/ . The unit cost of P_2O_5 is

    (a) Rs 24   (b) Rs 54   (c) Rs 54   (d) Rs 49

KEY TO EXERSISE - 6

1. b	2. a	3. b	4. b	5. a	6. c
7. a	8. a	9. a	10. d	11. d	12. b
13. c	14. c	15. c	16. d	17. b	18. c
19. a	20. d	21. b	22. a	23. a	24. c
25. d					

EXCERSISE - 7

1. Glucose and Fructose are
   - (a) Enantiomers
   - (b) Functional isomers
   - (c) Metamers
   - (d) Tautomers

2. Nicotine occurs mainly in
   - (a) Leaves
   - (b) Root
   - (c) Stem
   - (d) Flower

3. The isomer that is responsible for insecticidal activity of lindane is
   - (a) Alfa
   - (b) Beta
   - (c) Gamma
   - (d) Delta

4. Fertilizer which contains nitrogen in organic form is
   - (a) Ammonium nitrate
   - (b) Sodium nitrate
   - (c) Ammonium sulphate
   - (d) Urea

5. A fungicide which is highly toxic to fish is
   - (a) Carboxin
   - (b) Calixin
   - (c) Carbendazim
   - (d) Copper oxy chloride

6. The most abundant organic compound in nature is
   - (a) Cellulose
   - (b) Chitin
   - (c) Glucose
   - (d) Glycogen

7. Rock phosphate is suitable for application to long duration crops in
   - (a) Saline soils
   - (b) Sodic soils
   - (c) Acid soils
   - (d) Calcareous soils

8. Muriate of potash is not recommended for
   - (a) Rice
   - (b) Sorghum
   - (c) Sugar cane
   - (d) Sun flower

9. An example for rodenticide is
   - (a) Rotenone
   - (b) Zinc phosphide
   - (c) Dicofol
   - (d) DDT

10. Bio fertilizer used to fix nitrogen in jowar crop is
    - (a) Nitrobacter
    - (b) Rhizobium
    - (c) Azatobacter
    - (d) Rhodo spirillum

11. Fertilizer use efficiency of applied P fertilizer is
    - (a) 5-10%
    - (b) 10-25%
    - (c) 45-60%
    - (d) 30-45%

12. A herbicide which effectively controls Cyprus species of weeds is
    - (a) Atrazine
    - (b) Simazine
    - (c) Glyphosate
    - (d) Paraquat

13. An example of leguminous crop which is used for manure, fodder and pulse purpose is
    - (a) Sunnhemp
    - (b) Pillipesara
    - (c) Daincha
    - (d) Indigo

14. The water soluble P_2O_5 content of DAP is
    - (a) 46%
    - (b) 41%
    - (c) 18%
    - (d) 45%

15. Maximum permissible limit of biuret content in urea is

    (a) 2%   (b) 1.5%   (c) 2.5%   (d) 3.5%

16. Guano is a product obtained from

    (a) Fish   (b) Sea birds   (c) Cattle   (d) Sheep

17. The combustable gas in biogas plant

    (a) Methane   (b) Butane   (c) Ethane   (d) Pentane

18. A fungicide which also acts as a micro nutrient with in the plants is

    (a) Captan   (b) Carboxin   (c) Thiodan   (d) Zineb

19. The central P atom of organo phosphorus insecticides phosphorylates an enzyme

    (a) Oxidase   (b) Choline esterase

    (c) Carboxylase   (d) Sulfoxidase

KEY TO EXERSISE - 7

1. b	2. a	3. c	4. d	5. b	6. a
7. c	8. c	9. b	10. c	11. b	12. c
13. b	14. b	15. b	16. b	17. a	18. d
19. b					

UNDERLINE THE CORRECT ANSWER

1. The relative distribution of primary soil particles in a given soil is known as

    (a) Soil structure   (b) Soil texture

    (c) Soil plasticity   (d) Soil consistency

2. The ratio of volume of air to volume of soil is termed as

    (a) Aeration porosity   (b) Porosity

    (c) Air filled porosity   (d) None of these

3. The particle density of mineral soil is normally in the range of

    (a) 1.3-1.5 Mg/m^3 (b) 2.8-3.1 Mg/m^3

    (c) 2.6-2.75 Mg/m^3 (d) 2.0-2.4 Mg/m^3

4. Soil moisture condition at friable consistency is

    (a) Dry   (b) Moist   (c) Wet   (d) Saturated

5. Evaporation is an /a
   (a) Endothermic reaction   (b) Exothermic reaction
   (c) Hyperthermic reaction   (d) Isothermic reaction

6. The soil moisture tension at pF value 4.2 is
   (a) Field capacity   (b) Wilting point
   (c) Hygroscopic coefficient   (d) Air dry soil

7. Available water in soil refers to that water which is held between
   (a) Field capacity and hygroscopic coefficient
   (b) A maximum water holding capacity and field capacity
   (c) Wilting point and hygroscopic coefficient
   (d) Field capacity and permanent wilting point

8. Fast neutrons released from neutron moisture meter are effectively thermalized in soil by nuclei of
   (a) Oxygen   (b) Hydrogen   (c) Carbon i   (d) Helium

9. Matric potential is due to
   (a) change in pressure of air
   (b) Gravitational forces
   (c) Osmotic force
   (d) Adsorptive and capillary forces

10. In unsaturated condition the flow of water among the following is maximum in
    (a) Silty clay   (b) Clay loam   (c) Sandy loam (d) Silty loam

11. Tensiometer practically measures soil water potential up to
    (a) –1 bar   (b) —5 bar   (c) –0.5 bar   (d) -0.8 bar

12. Almost all natural ground water motion has Reynold's number
    (a) < 1   (b) 1-10   (c) 10-100   (d) >100

13. Maximum potential difference in SPAC occurs between
    (a) Soil and root   (b) Root and stem
    (c) Stem and leaf   (d) Leaf and external atmosphere

14. The carbon dioxide content of soil air is about
    (a) 0.03%   (b) 0.25%   (c) 2.5%   (d) 3.0%

15. For normal growth of plants and soil organisms the ODR values should be at least

    (a) $80 \times 10^{-8} g/cm^2\text{-}min$ (b) $50 \times 10^{-8} g/cm^2\text{-}min$

    (c) $40 \times 10^{-8} g/cm^2\text{-}min$ (d) $10 \times 10^{-8} g/cm^2\text{-}min$

16. Specific heat of dry mineral soil is

    (a) < 1   (b) >1   (c) equal to 1   (d) None of these

17. Which one of the following would be the most important factor in influencing soil temperature?

    (a) Humus   (b) Water   (c) Ca content   (d) Fe content

18. The colour of the least productive soil is

    (a) White   (b) Gray   (c) Red   (d) Black

19. Fourier's law deals with

    (a) Water flow   (b) Air flow   (c) Mass flow   (d) Heat flow

20. Density of water is maximum at

    (a) $0\ ^\circ C$ (b) $4\ ^\circ C$ (c) $1\ ^\circ C$ (d) both a and b

21. Aggregation in soils is improved by using fertilizer containing

    (a) Phosphorus   (b) Micro nutrient   (c) Potassium   (d) Nitrogen

22. Modulus of rupture test is to know the

    (a) Strength of the soil   (b) Structure of soil

    (c) Texture of soil   (d) All the above

23. The saturated hydraulic conductivity of the soil is in the following order

    (a) Sand <Loam>clay   (b) Loam > Sand>Clay

    (c) Sand> Loam >Clay   (d) Loam <Sand> Clay

24. The component of soil air which influence the availability of nutrients to plants is

    (a) Oxygen   (b) Carbondioxide   (c) Nitrogen   (d) Water vapour

---

KEY TO UNDERLINE THE CORRECT ANSWER

1. b	2. c	3. c	4. b	5. a	6. b
7. d	8. b	9. d	10. a	11. d	12. a
13. d	14. b	15. c	16. a	17. b	18. a
19. d	20. b	21. a	22. a	23. c	24. b

EXCERSISE - 8

FILL IN THE BLANKS

1. Study of soils from stand point of higher plants is known as ________________

2. The smallest geometrical position of the crystal, which can be used as the repetitive unit to build up crystal is called as ________________

3. Mixing up of soil material in the solum by churning process caused due to swelling and shrinking of clays is known as ____________

4. The clay mineral having highest CEC is ____________

5. The clay mineral having highest layer charge is____________

6. The smallest volume that can be called "a soil " is ______________

7. The master horizon with least biological activity is ___________

8. The land capability sub-class indicates ________________

9. The kind of map used in detailed soil survey is ________________

10. Cryoturbation is observed in soils belonging to____________ soil order

UNDERLINE THE CORRECT ANSWER

1. The clay mineral whose peaks don't collapse after heating to 550 ^{0}C is
   (a) Smectite   (b) Illite   (c) Kaolinite   (d) Halloysite

2. Aging of soils is more rapid in the following climates
   (a) Warm – humid   (b) Cool-humid
   (c) Cold and hot   (d) Warm -cool

3. The process of transformation of raw organic matter into humus is known as
   (a) Podzolization   (b) Laterization
   (c) Humification   (d) Melanization

4. The poly pedon is also known as
   (a) Soil individual   (b) Pedon
   (c) Profile   (d) Control section

5. The lowest category in soil taxonomy is
   (a) Series   (b) Order   (c) Great group(d) Sub -group

6. The dominant soil group in India is

    (a) Black soil   (b) Red soil   (c) Alluvial soil (d) Saline soil

7. Man made diagnostic horizon

    (a) Moliic   (b) Plaggan   (c) Cambic   (d) Umbric

8. The first soil map of India was prepared by

    (a) Govinda Rajan (b) R.S.Murthy   (c) J.L.Sehgal   (d) Schokalaskya

9. The bas map used in reconnaissance soil survey

    (a) Toposheet   (b) Cadestral map

    (c) Atlas map   (d) Aerial photographs

10. Highly weathered soils belongs to the following order

    (a) Ultisols   (b) Oxisols   (c) Ardisols   (d) Inceptisols

TRUE OR FALSE

1.   Both amorphous and crystalline minerals were identified by DTA method
      [   ]

2.   Solum + parent material is termed as regolith   [   ]

3.   Podzolization occurs under tropical climate with basic parent material
      [   ]

4.   The Vermiculite is also known as $14 A^0$ mineral [   ]

5.   The epipedon with > 50 % base saturation is Mollic [   ]

Given below are the statement containing two aspects. Encircle the correct answer

1.   The dominate soil type in India is Alluvial soils .The alluvial soils are
      highly fertile

      TT/TF/FT/ FF

2.   Soils with andic properties were classified in to Andisols. They are found
      in plolar and circumpolar region.

      TT/TF/FT/ FF

3.   Vertic Haplustepts is a category of Vertisols and its category is sub-group

      TT/TF/FT/ FF

4.   Topography refers to difference in elevation .It is an active soil forming
      factor

      TT/TF/FT/ FF

5. Smectite is a 2:1 mineral and source of negative charge is isomorphous substitution.

   TT/TF/FT/ FF

KEY TO FILL IN THE BLANKS

1. Edaphology
2. Unit cell
3. Argillopedoturbation
4. Vermiculite
5. Illite / Hydrous mica/ Fine grained micas
6. Pedon
7. C-horizon
8. Dominant limitation
9. Cadestral map
10. Gelisols

KEY TO UNDERLINE THE CORRECT ANSWER

1. (b) Illite
2. (a) Warm-humid
3. (c) Humification
4. (a) Soil individual
5. (a) Series
6. (c) Alluvial
7. (b) Plaggan
8. (d) Schokalaskya
9. (a) Toposheet
10. (b) Oxisol

KEY TO TRUE OR FALSE

1. T   2. T   3. F   4. T   5. T

Key to Given below are the statements containing two aspects. Encircle the correct answer

1. TT   2. TF   3. FT   4. TF   5. TT

EXCERSISE - 9 (MOCKTEST - 1)

1. The factor responsible for acid soil is-
   (a) Low rainfall
   (b) high rainfall
   (c) high evaporation
   (d) Irrigation with saline water

2. High salt tolerant crop is-
   (a) Barley
   (b) Cotton
   (c) Maize
   (d) Pulses

3. Saline soils are those soils which have-
   (a) ECe > 4 mmohs/cm,pH >8.5, ESP > 15
   (b) ECe > 4 mmohs/cm, pH < 8.5, ESP > 15
   (c) ECe > 4 mmohs/cm,pH <8.5, ESP < 15,SAR < 13
   (d) ECe < 4 mmohs/cm, pH < 8.5, ESP > 15

4. SAR is equal to-
   (a) $Na+/(Ca^{2+} + Mg^{2+}/2)^{1/2}$
   (b) $Na+/(Ca^{2+} + Mg^{2+})^{1/2}/2$
   (c) $(Ca^{2+} + Mg^{2+}/Na^{+})^{1/2}$
   (d) $Na+/(Ca^{2+} + Mg^{2+})^{1/2}$

5. The harmful effect of saline soil is due to-
   (a) high osmotic pressure of soil solution
   (b) Low osmotic pressure of soil solution
   (c) high pH
   (d) Low pH

6. Which of the following is correct explanation of ESP-
   (a) Exchangable Na/CEC
   (b) Exchangable Na * 100/CEC
   (c) Exchangable Na * 100
   (d) CEC * 100/Exchangable Na

7. The saline soil can be managed by-
   (a) Addition of gypsum
   (b) Leaching of salts
   (c) Addition of lime
   (d) Addition of pyrite

8. The acidity of H+ in aquous phase of soils is called-
   (a) Reserve acidity
   (b) Active acidity
   (c) Permanent charge
   (d) None of the above

9. Lime is added to reclaim-
   (a) Alkali soil
   (b) Acid soil
   (c) Saline soil
   (d) Laterite soil

10. Which of the following characterstics helps in identification of alkali soil-
    (a) Black spots or colour on the surface
    (b) White or ash coloured layer of salts on the surface
    (c) Reddish colour on surface
    (d) None of the above

11. ZnSO4 is generally added to take more production under-
    (a) Acid soil   (b) Saline soil
    (c) Waterlogged soil   (d) Alkali soil

12. Total acid soil found in India-
    (a) 26 Mha   (b) 23 Mha   (c) 49 Mha   (d) 50 Mha

13. Which if the following equation is correct-
    (a) ECC = CCE * fineness factor   (b) fineness factor * ECC = CCE
    (c) NI = CCE * fineness factor   (d) both a & c

14. Due to overliming-
    (a) Deficiency of Fe ,Cu , & Zn will occur
    (b) P & K availability will be reduced
    (c) Scab in root crops will increase   (d) All

15. What happens when an saline – alkali soil is leached with water-
    (a) ph & EC both decreases   (b) pH increases
    (c) pH & EC both increases   (d) pH & EC both remains same

16. Which is correct in increasing order of acid tolerance.
    (a) Blueberry > Tobacco > Cotton > Sugarbeet
    (b) Blueberry < Tobacco < Cotton < Sugarbeet
    (c) Blueberry < Cotton < Tobacco < Sugarbeet
    (d) Sugarbeet > Cotton >Blueberry >Tobacco

17. Clubroot disease of cabbage may be controlled by raising soil pH to -
    (a) 5 & above   (b) 6 & above   (c) 7 & above   (d) 8 & above

18. Each unit change in pH makes a ten fold change in activity of -
    (a) H+ ions   (b) OH- ions
    (c) Both H+ & OH- ions   (d) none

19. Which of the following equation is correct.
    (a) $Eh = Eo - RT/nF \ln (ox/red)$ (b) $Eh = Eo + RT/nF \ln (ox/red)$
    (c) $Eh = Eo + RT/nF \ln (red/ox)$ (d) none

20. In submerged soil –
    (a) specific conductance of solution decreases
    (b) specific conductance of solution increases
    (c) at first increases, attain a maximum and declines to a fairly stable value
    (d) None

21. Which of the following equation is correct-
    (a) $pE = Eh/0.0591$ (b) $pE = Eh * 0.0591$
    (c) $pE = -\log (e)$ (d) Both a & c

22. The solubility of native from of Zn after submergence
    (a) decrease by a factor 102 for each unit increase in soil pH
    (b) increase by a factor 102 for each unit increase in soil pH
    (c) decrease by a factor 10 for each unit increase in soil pH
    (d) none

23. Reduction of Mn4+ occurs in submerged soil when redox potential value is-
    (a) +150 mV to +300mV   (b) +200 mV to +450mV
    (c) +200 mV to +300mV   (d) +200 mV to +400mV

24. Soils are prone to puddling when clay particles contain-
    (a) > 20%   (b) < 20   (c) > 15%   (d) > 18%

25. Submerged soils are-
    (a) a partially oxydised A horizon high in O.M
    (b) A mottled zone
    (c) a partially reduced zone   (d) All of the above

26. Acid soils have been classified by USDA –
    (a) pH < 6.5 in 1 : 1soil – water suspension
    (b) pH < 5.5 in 1 : 1soil – water suspension
    (c) pH < 5.5 in 2 : 1soil – water suspension
    (d) pH < 6.5 in 2 : 1soil – water suspension

27. Composition of SMP buffer is –
    (a) 1.8 gm nitrophenol, 2.5 ml of triethanolamine
    (b) 1.8 gm nitrophenol, water pH 8.5
    (c) 1.8 gm nitrophenol, $CaCl_2$ pH 8.5
    (d) 1.8 gm nitrophenol, 2.5 ml of triethanolamine, 3 gm potassium chromate, 2 gm Calcium acetate & 53.2 gm CaCl2 dihydrate pH 7.5 with dilute NaOH.

28. Kari & pokali soils are found in-
    (a) Karnataka   (b) Maharastra
    (c) Kerala   (d) Tamilnadu

29. Gypsum factor is-
    (a) 1.2 to 1.5   (b) 1.0 to 1.5   (c) 2.5 to 3.0   (d)1.2 to 1.7

30. Sodic soils in india are dominant in areas with a mean annual rainfall-
    (a) 50 to 90 cm   (b) > 90 cm
    (c) 55 to 90 cm   (d) < 55.

31. LR is

    (a) (Eq. Depth of drainage water * 100)/ Eq. Depth of applied water

    (b) ECaw * 100 /ECdw

    (c) ECaw/ECdw

    (d) Both a & b

32. To reclaim saline soils-

    (a) continuous ponding is better than intermittent ponding

    (b) Intermittent ponding is better than continuous ponding

    (c) both a & b are correct

    (d) none

33. Mn toxicity is found in acid soils when

    (a) pH < 4.0   (b) < 3.0   (c) < 3.5   (d) < 5

34. For a sodic soil with CEC = 10 & ESP = 40, gypsum requirement to lower the ESP to 10 will be-

    (a) 5 mmol of Ca2+ per 100 gm of soil

    (b) 8 mmol of Ca2+ per 100 gm of soil

    (c) 3 mmol of Ca2+ per 100 gm of soil

    (d) 4 mmol of Ca2+ per 100 gm of soil

35. Iron pyrite abundantly available in-

    (a) Bihar   (b) W. B   (c) Assam   (d) U. P

36. Which of these equation is correct-

    (a) ESP =100 * (-0.0126 + 0.01475 SAR)/ 1 + (0.0126 + 0.01475 SAR)

    (b) ESP/100 = (-0.0126 + 0.01475 SAR)/ 1 + (0.0126 + 0.01475 SAR)

    (c) ESP/100 = SAR (-0.0126 + 0.01475)/ 1 + (0.0126 + 0.01475 SAR)

    (d) both a & b

37. Magnesium hazard in irrigation water is expected having Mg : Ca ratio –

    (a) equal to 1   (b) Less than 1   (c) More than 1 (d) None

38. High sodium hazard has SAR value in between-

    (a) 10 – 18   (b) > 26   (c) 18 – 26   (d) None

39. The leaching requirement of an irrigation water having EC of 3 dS/m and EC of drainage water is 8 dS/m will be-

    (a) 37%   (b) 38.5%   (c) 37.5%   (d) 40%

40. Langeliar's saturation index is-
    (a) $1 + (8.4 - pHc)$
    (b) $(8.4 - pHc)$
    (c) $1 - (8.4 + pHc)$
    (d) None

41. Which of the following equation is correct-
    (a) RSC (me/l)= $(CO_3^{2-} - HCO_3^-) + (Ca^{2+} - Mg^{2+})$
    (b) RSC (me/l)= $(CO_3^{2-} + HCO_3^-) + (Ca^{2+} + Mg^{2+})$
    (c) RSC (me/l)= $(CO_3^{2-} + HCO_3^-) - (Ca^{2+} + Mg^{2+})$
    (d) RSC (me/l)= $(CO_3^{2-} + HCO_3^-) - (Ca^{2+} - Mg^{2+})$

42. The salt index for irrigation water of high quality is-
    (a) positive
    (b) Negative
    (c) May be positive or negative
    (d) none

43. Water can be used with certain management should have RSC value between-
    (a) 0.5 to 1
    (b) 1.25 to 2.5
    (c) 2.5 to 2.75
    (d) 1 to 1.25

44. High hazard to water quality when boron concentration is –
    (a) 1 to 2ppm
    (b) 2 to 4ppm
    (c) < 1 ppm
    (d) None

45. Liming factor is generally-
    (a) 2 to 2.5
    (b) 2.5 to 3.0
    (c) 1 to 1.5
    (d) 1.5 to 2.0

MATCH THE FOLLOWING

1.   AMENDMENT   **GYPSUM EQUIVALENT**

(a)	Gypsum	i.	1.0
(b)	Iron sulphate	ii.	1.62
(c)	Aluminium sulphate	iii.	1.29
(d)	Calcium polysulphide	iv.	0.77
(e)	Iron pyrite	v.	0.6

2. Salinity scale   **Plant response**

(a)	0 – 2	i.	Tolerant crops yield satisfactorily
(b)	4 – 8	ii.	Very tolerant crop yield satisfactorily
(c)	> 16	iii.	Yields of crops restricted
(d)	2 – 4	iv.	Negligible effect
(e)	8 – 16	v.	yields of very sensitive crops restricted

3. **Reduction**   **Redox potential**
   (a) O_2 to H_2O i.   -200 to -280
   (b) NO_3 to N_2 ii.   -120 to -180
   (c) Fe^{3+} to Fe^{2+} iii. + 180 to +150
   (d) SO_4^{2-} to S^- iv.   +380 to +320
   (e) CO_2 to CH_4 v.   +280 to +220

4. **Property**   **For irrigation purpose**
   (a) BOD   i.   500 mg/l
   (b) COD   ii.   100 mg/l
   (c) TSS   iii. 200 mg/l
   (d) TDS   iv.   2100 mg/l

5. **Liming material**   **CCE**
   (a) $Ca(OH)_2$ i.   108.7
   (b) CaO   ii.   86
   (c) $CaCO_3 . MgCO_3$ iii. 100
   (d) $CaCO_3$ iv.   136
   (e) $CaSio_3$ v.   179

EXCERSISE - 10 (MOCKTEST - 2)

1. Study of origin, classification and description of soil is known as
   - (a) Petrology
   - (b) Petrography
   - (c) Pedography
   - (d) Pedology

2. Vertical section through a soil is called
   - (a) Soil horizon
   - (b) Soil solum
   - (c) Soil profile
   - (d) Regolith

3. _________ horizon is called Eluvial horizon (zone of washing out) ?
   - (a) A
   - (b) B
   - (c) C
   - (d) O
   - (e) none of these

4. _________ horizon is called Illuvial horizon ?
   - (a) A
   - (b) B
   - (c) C
   - (d) O

5. Solum of a soil profile is
   - (a) B+C horizons
   - (b) O+A horizons
   - (c) A+B horizons
   - (d) A+B+C horizons

6. Regolith is
   - (a) O horizon
   - (b) B+C horizon
   - (c) A+B+ partially weathered C horizon
   - (d) A+ C horizon

7. Granite is an example for
   - (a) Intrusive igneous rock
   - (b) Extrusive igneous rock
   - (c) Sedimentary rock
   - (d) Metamorphic rock

8. Stratification is a characteristic feature of ________ rocks
   - (a) Igneous rocks
   - (b) Sedimentary rocks
   - (c) Metamorphic rocks
   - (d) both a & b

9. Most important chemical weathering process is
   - (a) Hydration
   - (b) Hydrolysis
   - (c) Carbonation
   - (d) Oxidation & reduction

10. Mechanical analysis (process of seperation of soil fractions) is based on
    - (a) Fick's law
    - (b) Darcy's law
    - (c) Stocke's law
    - (d) Fourier's law

11. A soil is classified as clayey if it contains atleast ________ % clay
    - (a) 15
    - (b) 35
    - (c) 70
    - (d) 50

12. The relative proportion of sand, silt and clay in a soil is known as
    - (a) Soil structure
    - (b) Soil texture
    - (c) Plasticity
    - (d) Bulk density

13. Mass of unit volume of soil including the pore space is called
    (a) Particle density
    (b) Bulk density
    (c) Apparent density
    (d) both b&c

14. Natural soil aggregates are called
    (a) Clod
    (b) Ped
    (c) Fragment
    (d) Concretion

15. CO_2 content of soil is
    (a) 0.03 %
    (b) 0.5 %
    (c) 3 %
    (d) 30 %

16. Dominant spectral colour in Munsell colour notation is represented by
    (a) Hue
    (b) Value
    (c) Chroma
    (d) all of these

17. Logarithm of centimeter height of a water column to give the necessary suction is known as
    (a) p^H scale
    (b) pF scale
    (c) pG scale
    (d) Arrhenius scale

18. Soil water retained between -0.33 bar & -15 bars is called
    (a) Capillary water
    (b) Available water
    (c) Hygroscopic water
    (d) Gravitational water

19. Tensiometer and Neutrone moisture meter is used for measurement of
    (a) Soil evaporation
    (b) Soil moisture
    (c) Soil temperature
    (d) Soil organic matter content

20. Movement of water into the soil is called __________
    (a) Infiltration
    (b) Percolation
    (c) Tortuosity
    (d)Saturated flow

21. Amount of water required to produce a unit quantity of dry weight material is called
    (a) Water use efficiency
    (b) Consumptive use
    (c) Transpiration ratio
    (d) Biomass production

22. Capacity of the soil to change its shape under moist conditions is called
    (a) Interference
    (b) Adhesion
    (c) Plasticity
    (d) Elasticity
    (e) none of these

23. Kaolinite is an example for
    (a) 1:1 type mineral
    (b) 2:1 expanding mineral
    (c)
    (d) 2:1 non-expanding mineral
    (e) 2:2 type

24. Montmorillonite is an example for
    (a) 1:1 type
    (b) 2:1 expanding
    (c) 2:1 non-expanding
    (d) 2:2

25. The mineral which is not available in strongly acidic soils
    (a) Iron
    (b) Zinc
    (c) Molybdenum
    (d) Copper
    (e) Manganese

26. Fungi prefers
    (a) Alkaline conditions
    (b) Acidic conditions
    (c) Neutral conditions
    (d) both a & c
    (e) none

27. The power to resist a change in pH by the soil is called its
    (a) Plasticity
    (b) Buffering capacity
    (c) Healing capacity
    (d) Moulding capacity

28. Browning disease in rice is caused by the deficiency of
    (a) Mn
    (b) Fe
    (c) Zn
    (d) both a & b

29. Which one of the following is not a liming material ?
    (a) Quick lime
    (b) Slaked lime
    (c) Basic slag
    (d) calcite
    (e) none

30. The process of removal of silica instead of sessquioxides to concentrate in the solum is called
    (a) Podzolisation
    (b) Laterisation
    (c) Brucification
    (d) both a & c

31. Re-precipitation of sesquioxides and humification of organic material is seen in
    (a) Laterisation
    (b) Podzolisation
    (c) Ferruginisation
    (d) both b & c

32. Saline soils are also called
    (a) Solonchak
    (b) White alkali soils
    (c) Black alkali soils
    (d) both a & b

33. Alkali soils are also called
    (a) Sodic soils
    (b) non-saline alkali soils
    (c) Solonetz
    (d) a, b & c

34. Example for a highly salt resistant crop is

    (a) Barley   (b) Beans   (c) Radish   (d) a, b & c

35. Severe erosion at some hilly areas may erode the surface A horizons leaving
    B & C horizons only. These soils are known as

    (a) Truncated soils   (b) Elluded soils

    (c) Illuded soils   (d) both a & c

36. In nitrification process, conversion of NH_3 to nitrite is done by

    (a) Nitrobacter   (b) Nitrosomonas

    (c) Clostridium   (d) Azotobacter

37. Conversion of nitrite to nitrate is done by

    (a) Nitrobacter   (b) Nitrosomonas

    (c) Clostridium   (d) Azotobacter

38. Example of a non symbiotic nitrogen fixer is

    (a) Clostridium   (b) Azotobacter

    (c) Azospirillum   (d) all of these

39. Potato scab is controlled by the application of

    (a) Bordeaux mixture   (b) Sulphur

    (c) Gypsum   (d) NaCl

40. National Bureau of Soil Survey and Land Use Planning (NBLSS & LUP)
    is located at

    (a) Bhopal   (b) Nagpur   (c) Kanpur   (d) Izathnagar

41. Number of Land suitability classes is

    (a) 8   (b) 4   (c) 2   (d) 12

42. Soil class suitable for wild life conservation

    (a) I   (b) IV   (c) VI   (d) VIII

43. Total number of soil irrigability classes

    (a) 4   (b) 6   (c) 8   (d) 5

44. Largest and most important soil group of India contributing the maximum
    share to its agricultural wealth

    (a) Red soil   (b) Black soils

    (c) Alluvial soils   (d) Laterite soils

45. The nutrients which are known as secondary macro nutrients are

    (a) N,P,K   (b) N,P,S   (c) Ca,Mg,S   (d) Mo,Fe,S

46. Plants can store _____ element in larger quantities than what is needed for optimum growth, without causing toxicity (Luxury consumption)

    (a) N   (b) Mg   (c) P   (d) K   (e) B

47. Formation of starch and translocation of sugars are important functions of

    (a) N   (b) P   (c) K   (d) Ca

48. Which of the following element is involved in the synthesis of IAA

    (a) Ca   (b) Fe   (c) B   (d) Zn

49. Which element is essential for pollen germination and growth of pollen tube ?

    (a) Cu   (b) B   (c) Mo   (d) Cl

50. Criteria of essentiality of nutrients was proposed by

    (a) Cate & Nelson

    (b) Arnon & Stout

    (c) Bacon & Jacobs

    (d) Subbaiah & Assija

FILL IN THE BLANKS

1. The pH at which the net surface charge of the soil colloids is zero is known as —————————

2. The deficit of anions near a clay surface when compared to equilibrium solution is known as —————

3. The conc of Na^+ in soil solution decreases four times on dilution, then the conc of Ca^{+2} must also decreases ————————— times

4. The irreversibility in K^+ ionic fixation and release process in illitic mineral is due to —————

5. Substitution of lower valent cation by a higher valent ion in a structral lattice leads to —— charge on clay surface

6. Sodium saturation in soil is expressed as ————————— and —— —————————

7. The crystals of Kaolinite are joined by ————————— bonding making the mineral a non expanding

8. Saline soils can be reclaimed by the process of—————————

9. The relationship between the amount adsorbed per unit mass of the adsorbent and the equilibrium conc of the adsorbate at a constant temperature is shown in the form of —————

10. The pH at which the net surface charge of the soil colloids is zero is known as ————————

11. The deficit of anions near a clay surface when compared to equilibrium solution is known as ————————

12. The conc of $Na+$ in soil solution decreases four times on dilution, then the conc of $Ca+2$ must also decreases ———————— times

13. The irreversibility in $K+$ ionic fixation and release process in illitic mineral is due to ————————

14. Substitution of lower valent cation by a higher valent ion in a structral lattice leads to —— charge on clay surface

15. Sodium saturation in soil is expressed as ———————— and ——
————————————

16. The crystals of Kaolinite are joined by ———————— bonding making the mineral a non expanding

17. Saline soils can be reclaimed by the process of ————————

18. The relationship between the amount adsorbed per unit mass of the adsorbent and the equilibrium concentration of the adsorbate at a constant temperature is shown in the form of ————————